Crashworthiness of Composite Thin-Walled Structural Components

Crashworthiness of Composite Thin-Walled Structural Components

A. G. MAMALIS, Dr.-Ing., Professor
D. E. MANOLAKOS. Dr.-Ing., Assistant Professor
G. A. DEMOSTHENOUS, Dr.-Ing.
M. B. IOANNIDIS, Dr.-Ing.

Manufacturing Technology Division
Department of Mechanical Engineering
National Technical University of Athens
Greece

CRC Press
Taylor & Francis Group
Boca Raton London New York

CRC Press is an imprint of the
Taylor & Francis Group, an **informa** business

CRC Press
Taylor & Francis Group
6000 Broken Sound Parkway NW, Suite 300
Boca Raton, FL 33487-2742

First issued in paperback 2019

ISBN-13: 978-1-56676-635-7 (hbk)
ISBN-13: 978-0-367-40035-4 (pbk)

Library of Congress Cataloging-in-Publication Data

Main entry under title:
Crashworthiness of Composite thin-Walled Structural Components

Library of Congress Card Number 97-62362

Visit the Taylor & Francis Web site at
http://www.taylorandfrancis.com

and the CRC Press Web site at
http://www.crcpress.com

TABLE OF CONTENTS

PREFACE

The introduction which follows sufficiently describes our aims in presenting this monograph on the crashworthy behaviour of thin-walled structural components of fibreglass composite materials which are subjected to various loading conditions, relevant to vehicle design and manufacture. Essentially it comprises the results of our extensive theoretical and experimental work on the topic and the material of a few lectures on Crashworthiness and failure mechanisms of composites given to undergraduates in Mechanical Engineering. The monograph is intended to illustrate and indicate the engineering design outlets and applications of the analytical work, mainly on fracture and failure, and new notions and considerations in vehicle engineering situations. We hope that the contents of our work will be of value to students, teachers and many kinds of professional engineers.

PROFESSOR DR.- ING. A.G. MAMALIS
ASSISTANT PROF. DR.- ING. D.E. MANOLAKOS
DR.- ING. G.A. DEMOSTHENOUS
DR.- ING. M.B. IOANNIDIS

CHAPTER 1

INTRODUCTION

Vehicle crashworthiness has been improving in recent years with attention mainly directed towards reducing the impact of the crash on the passengers. Effort has been spent in experimental research and in establishing safe theoretical design criteria on the mechanics of crumpling, providing to the engineers the ability to design vehicle structures so that the maximum amount of energy will dissipate while the material surrounding the passenger compartment is deformed, thus protecting the people inside.

The improvement of structural crashworthiness must be accomplished within certain constraints, such as limits on force transmission and/or deformation and failure. The mechanisms for improving vehicular resistance will depend on the nature of the imposed limitations. If only small deformations are permitted, then large amounts of momentum transfer and force levels to the occupant must be expected, which may be unacceptable. Conversely, if large permanent deflections are tolerable, a limit can be set on the magnitude of the force experienced by the occupant provided that the crush of the capsule still retains a minimum volume for survival. An optimal way that this can be achieved is by exclusive use of frame deformation, including applications of specific energy absorbing devices, such as strategically placed tubular elements. However, vehicle impact processes are very complex events where the simultaneous structural response of many different interacting units renders their behaviour only rarely susceptible to the usual detailed analysis associated with a single component.

Many of the mechanical devices and elements, made of metals, polymers and composite materials, are designed to absorb impact energy under axial crushing, bending and/or combined loading. An important requirement is that these structural members must be able to dissipate large amounts of energy by controlled collapse, in the event of a collision. Generally, the total energy dissipated depends upon the governing deformation phenomena of all or part of structural components of simple geometry, such as thin-walled tubes, cones, frames and sections. The energy absorbing capacity differs from one component to the next in a manner which depends upon the mode of deformation involved and the material used.

During the last decade the attention given to crashworthiness and crash energy management has been centred on composite structures. The main advantages of fibre reinforced composite materials over more conventional isotropic materials, are the very high specific strengths and specific stiffnesses which can be achieved. Moreover, with composites, the designer can vary the type of fibre, matrix and fibre orientation to produce composites with improved material properties. Besides the perspective of reduced weight, design flexibility and low fabrication costs composite materials offer a considerable potential for lightweight energy absorbing structures; these facts attract the attention of the automotive and aircraft industry owing to the increased use of composite materials in various applications, such as frame rails used in the apron construction of a car body and the subfloor of an aircraft, replacing the conventional materials used.

Previous investigations indicated that composite shells deform in a manner different than similar structural components made of conventional materials, i.e. metals and polymers, since microfailure modes, such as matrix cracking, delamination, fibre breakage etc., constitute the main failure modes of these collapsed structures. Therefore, this complex fracture mechanism renders difficulties to theoretically modelling the collapse behaviour of fibre-reinforced composite shells.

Extensive research work has been performed primarily on axial loading and bending of simple thin-walled composite structures. The effect of specimen geometry on the energy absorption capability was investigated by varying the cross-sectional dimensions, wall thickness and length of the shell. The effect of the type of composite material, laminate design, loading method and strain-rate on the crashworthy behaviour of the components was also studied. Environmental effects related to crash characteristics of composites have been also investigated.

In this monograph the crashworthy behaviour of thin-walled structural components of fibreglass composite materials, namely circular and square/rectangular tubes, circular and square frusta and automotive sections, subjected to various loading conditions, i.e. static and dynamic axial loading and bending, is considered. The loading and deformation characteristics of the collapsed shells are obtained experimentally and theoretically by modelling the crumpling and bending process; the influence of the shell geometry and the material properties on these characteristics, in relation to the behaviour of the shells as energy absorbing devices, is examined. Furthermore, lecturing material pertaining to failure mechanisms of the composite structures is introduced to provide the necessary notions regarding the behaviour of composite materials in vehicle engineering situations.

In particular, Chapter 2 of the monograph deals with the structural features relevant to vehicle collisions and, moreover, with the use of composite materials in crashworthiness applications. Chapter 3 mainly considers the failure mechanisms of composites, whilst in Chapter 4 an extensive review on energy absorption capability of thin-walled composite structural components is reported. Chapters 5, 6, 7, 8 and 9 deal with the axial collapse and bending of thin-walled structural components of fibreglass composite materials, i.e. circular tubes, square tubes, circular frusta, square frusta and automotive sections, both experimentally and theoretically. Finally, Chapter 10 usefully classifies the macro- and microfailure modes and the quantitative data obtained and reported in the previous chapters, providing, therefore, useful conclusions for vehicle design and manufacture.

Our monograph is intended to provide an introduction to this relatively new topic of structural Crashworthiness for professional engineers. It will introduce them to terms and concepts of it and acquaint them with some sources of literature about it. We believe that, our survey constitutes a reasonably well-balanced synopsis of the topic. It is also our hope to be writing for engineering students and teachers, to provide them with an exposure to the subject and, thus enhance interest in Crashworthiness.

CHAPTER 2

VEHICLE CRASHWORTHINESS

2.1 ASPECTS OF CRASHWORTHINESS

2.1.1 General

The techniques for providing direct human protection in vehicles are strongly related to the efficacy of the passenger cell to prevent damage to the occupant in vehicular collisions. Attention of the manufacturers, research and design engineers, bioengineers and other related professionals, has been directed towards it, in particular over the last thirty years. A large number of citations of publications of several annular conferences proceedings on automobile crush situations and design mechanics have been cited, whilst substantial efforts have been made by manufacturers to improve the crash resistance—generally termed Crashworthiness—of their vehicles. This concept is intimately related to the energy absorption capability of the vehicle. The combination of crashworthiness and the effectiveness of the body protector jointly determine the survivability of the occupants. Since there is a much more restricted capability of the latter to prevent human injury, it is evident that the vehicular resistivity must be enhanced to the maximum practical limit to provide the greatest level of safety. Reference [1] is a useful general starting point for all aspects of crashworthy impact situations.

Crashworthiness can be achieved in one of two ways depending upon (a) the minimum deformation of the exposed structural element or system, or (b) the maximum deformation up to a specified limit. In the former case, the frame of contents may be subjected to a high level of momentum and force transfer that may be unacceptable, whilst implicit in the second case is the consequence of large, irreversible displacements of a part or all of the structure, but the degree of force and momentum, trasmitted to the protective shell and its contents, is significantly reduced. An optimal way to fulfill the second goal is the use of suitable energy absorbing devices in strategical places of the vehicle.

The design aim is to dissipate kinetic energy irreversibly rather than convert and store it elastically and, in particular, restitution is to be avoided. Devices used to this end are usually one-shot items, i.e., once having been deformed, they are discarded and replaced. Frequently, they constitute a special kind of load-limiter, being proportioned so as to possess a more-or-less rectangular force-displacement characteristic. The cost of these devices must always be kept in mind and searches for high energy absorption-per-unit-weight or volume, which is very important in aircraft, may well be justified. However, after application, i.e., in post-traumatic collision situations, views about cost may well have changed.

Many of the mechanical devices and elements, designed to absorb impact energy under the conditions referred to above, generally depend upon the plastic deformation of all or part of common structural elements, such as tubes, frames, wires and bars. Note, that most of the cases are described with reference to their quasi-static characteristics, since these include the predominant geometrical effects, which occur under dynamic loading, whilst the effect of strain-rate in increasing the yield stress can be taken into account by using a simple scaling factor based on the mean strain-rate in the critical plastic zones. The implication of designing an energy-absorbing device on the basis of its quasi-static loading response is that inertia effects within the device itself are unimportant and, hence, the kinetic energy is considered converted into plastic work in a quasi-static deformation mode [1, 2].

Crashworthiness studies provide with the mechanism by which a proportion of impact energy is absorbed by the collapsing structure, whilst a small amount is transfered to the passenger. To obtain effective crashworthy behaviour, a crashworthiness study must be carried out in the very early design stages.

Two design procedures of the above mentioned analysis may be listed. The first is based on trial and error with testing prototype structures, whilst the second uses detailed mathematical models analysed by numerical simulation, e.g the finite element method. To apply this purely theoretical analysis in the initial stage of design, when detailed structural and material data are not available, is very difficult. On the other hand, the hand-made and impact testing of prototypes is very costly. For these reasons, the designer must be provided with a design tool capable to analyse alternative design solutions for various loading conditions.

The optimum design stages of a crashworthy structural energy absorbing system may be classified as, see Figure 2.1:

1. *Structural components data-base:* It constitutes the start of the design procedure. The component data are obtained through testing or analysis.
2. *Data acquisition of components by testing and analysis:* If the data base is impossible to be used for a particular design, the corresponding structural component is manufactured and tested. Its collapse mechanism is analysed so that failure of a certain part may be avoided by properly strengthening it. The results obtained are added to the data-base 1.
3. *Overall analysis of the collapse mechanism:* Following the evaluation of the non-linear properties of the structural components, regarding strength and energy absorption capability, a quasi-static analysis of the whole structure is carried out, in order to predict its total load carrying capacity. It is assumed, that the mass of the collapsed structure is negligible compared to the striking mass and the direc-

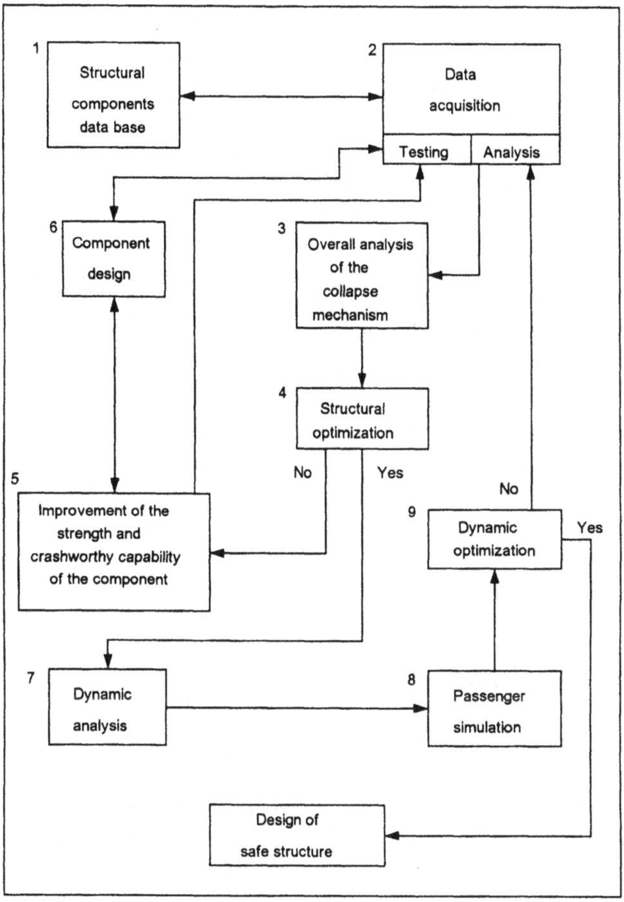

Figure 2.1. Flowchart of the optimum design procedure of energy absorbing systems.

tion of the loads exerted during impact on the collapsed structure may be predicted in advance.

4. *Structural optimisation:* Through the total load carrying capacity of the structure (Stage 3), the strength and energy absorption capability of each structural component are also evaluated. This information allows for the designer to re-design "incompatible" components. Criteria are imposed for evaluating the ability of the component to transfer the loads exerted throughout the impacted structure and to simultaneously absorb a significant amount of the dissipated energy, whilst high local deformations must be avoided.
5. *Improvement of the strength and crashworthy capability of the component:* Following the prediction of "incompatible" components, collapse analysis is carried out, in order to improve the strength and the energy absorption capability of the loaded components to fulfill the design requirements. Subsequently, the dimen-

sions of the improved "compatible" components are calculated, when, under compressive or bending loading, the necessary internal load is attained.

6. *Components design:* The simplified components predicted in Stage 5, which fulfill the crashworthiness requirements, are adequate for the initial design stages of the structure. However, due to many other requirements, the calculation procedure becomes complicated. Their contribution on the behaviour of the whole structure may be checked through Stage 3.
7. *Dynamic analysis:* Two possible cases may be associated with dynamic analysis: (a) If the mass of the collapsing structure is negligible compared to the striking mass and the impact velocity is not too high, i.e bus roll-over or car side-impact, the quasi-static load-deformation curve obtained in Stage 3 may be used. (b) If there are significant masses within the collapsing structure, the results obtained in Stages 1, 2, 5 and 6 may be used directly.
8. *Passenger simulation:* After completing Stage 7, the velocity and the collapsing structure are already known. In the present stage, it is necessary to consider the passenger or cargo safety during the crush, whilst Stage 4 ensures that loading and deformation are distributed to all structural elements, so that the optimum energy dissipation takes place. The effect of the overall (structure-passenger) impact on the passenger is analysed in detail.
9. *Dynamic optimisation:* If the results obtained in Stage 9 are not acceptable as far as the passenger safety is concerned, the whole analysis procedure described in Stages 1–8 must be repeated, until a safe structure is designed.

2.1.2 Structural Response to Impact

This topic has important and widespread application; the reader is referred to the monograph by Johnson and Mamalis [1] and the review by Rawlings [3], which contains a section specifically devoted to motor cars. Also, the review by Johnson and Reid [2], gives information about general features of dynamic structural response and about specific devices, which have been designed to absorb energy in impact situations.

The main types of vehicle impact have been tabulated by Franchini [4], see Figure 2.2. Head-on collision with barriers or between cars moving with the same speed, in the range 40–72 km/h, shows a time for retardation of about 0.1 s with an approximately linear rise to, and fall from, the greatest retardation of about 40 g, which occurs after about 0.05s; this involves a mean crushing strength of the car about 206 kN/m^2 of its minimum cross-sectional area, say 1.5 m^2.

PLASTIC COLLAPSE OF SHELLS AND STRUTS

Tubes provide perhaps the widest range of possible uses in energy absorbing systems of any simple structure. In addition to their use as energy absorbers, their common occurrence as structural elements implies an instrict energy-absorbing capability in certain vehicle structures. This dual role is most desirable, particularly in aircraft design where weight minimisation is important.

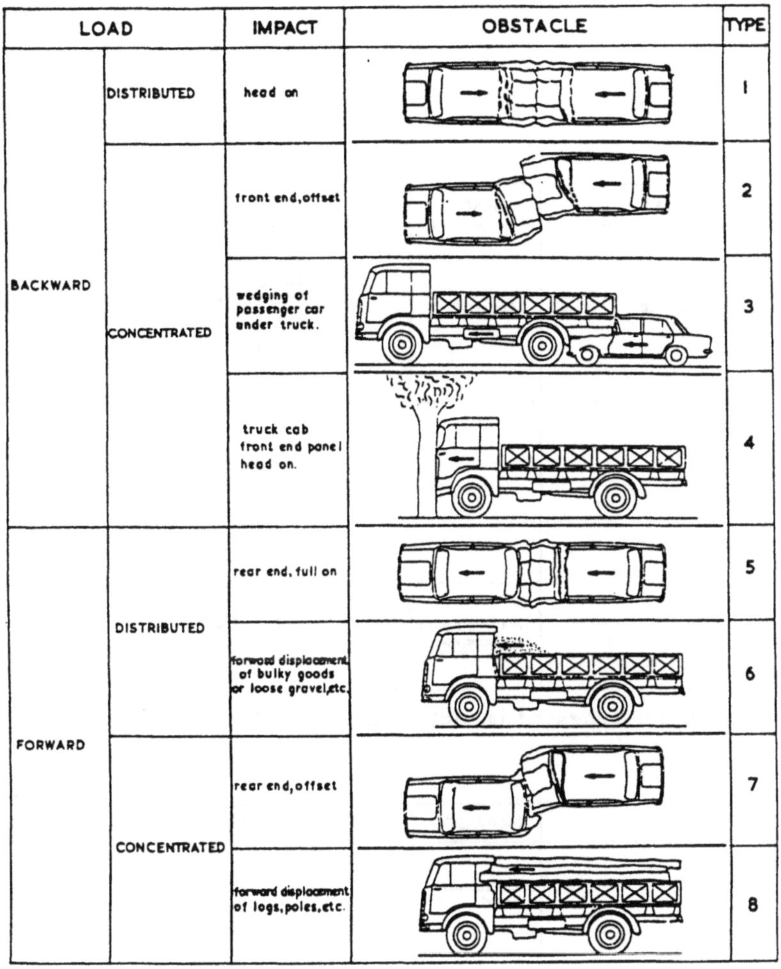

LOAD
IMPACT
OBSTACLE
TYPE
BACKWARD
DISTRIBUTED
head on
1
CONCENTRATED
front end, offset
2
wedging of passenger car under truck.
3
truck cab front end panel head on.
4
FORWARD
DISTRIBUTED
rear end, full on
5
forward displacement of bulky goods or loose gravel, etc.
6
CONCENTRATED
rear end, offset
7
forward displacement of logs, poles, etc.
8

(a) Main types of impact; longitudinal load

Figure 2.2. Main types of impact.

Shells made of ductile materials can be subjected to a wide range of deformation modes and various loading conditions, as discussed in the vast amount of literature dealing with lateral compression or tube flattening, local loading of tubes, axial buckling and bending of: circular cylindrical shells, square and rectangular tubes, conical shells and frusta, honeycomb material, tube inversion and sandwich plates. Furthermore, the buckling of solid struts was considered both experimentally and theoretically by many researchers. See the literature survey on these topics reported in References [1, 2, 5–8].

LOAD	IMPACT	OBSTACLE	TYPE
DISTRIBUTED	side, 90°	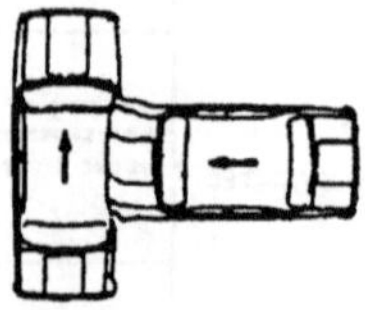	9
CONCENTRATED	side, oblique		10
	side skid against tree, pole, etc.		11

(b) Main types of impact; transverse load

LOAD	IMPACT	OBSTACLE	TYPE
DISTRIBUTED	full area, roof panel.		12
CONCENTRATED	side edge, roof panel.		13
	end edge, roof panel.		14

(c) Main types of impact; vertical load

Figure 2.2 (continued). Main types of impact.

ENERGY ABSORBING FRAMES

In Figure 2.3 a diagram of an energy absorbing S-frame which consists of heavy box members, is shown. Inset in the figure is the crash-front S-frame before and after a frontal impact. An elastic stress distribution for a large curvature beam was assumed for the design of this frame.

CRUMPLE ZONES

Many manufacturers of cars now design into their models regions at the front, behind the bumper, and rear of the car, which crumple during collision in a predetermined mode at a pre-selected rate, thereby aiming to minimize the retardation force, but preferably maintain it at a constant level which the passenger may have to bear, see Figure 2.4. Note that experimental facts indicate that, in the case of frontal or rear end collisions, the front and rear ends of the vehicle, respectively, collapse and absorb energy before its central portion deforms permanently, whilst approximately 90% of the energy goes into permanent plastic and 10% into elastic deformation [1]. Finite element analysis and computer-aided design are now commonly used for calculating the extent and degree of impact deformation.

BUMPERS

Bumper designs are manifold—these typically having curved overriders encased in rubber. Bumper mountings also include strong ribbed steel cones which concer-

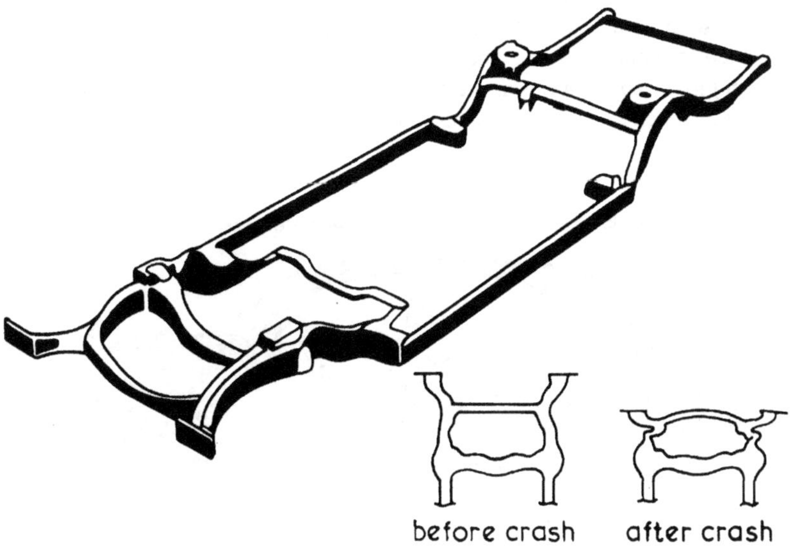

Figure 2.3. Energy absorbing frame.

tina in impact situations and hydraulic mountings are occasionally used. Bumpers are now often covered with resilient plastic layers so that they can also reduce injuries to pedestrians and cyclists, see References [1,7].

DRIVER AND PASSENGER COMPARTMENT

The "survival space" or a minimum residual volume after impact, which envelops occupants taking account of occupant size, driving posture, and seats is of almost constant magnitude regardless of car model. As stated in Reference [4], designed maintenance of survival space for collision speeds of at least up to 48 km/h is generally possible.

Strong steel compartments—cages or safety cell—sare now frequently designed into cars, see Figure 2.4; passengers are protected from above and below, between the front and rear bumpers and from the sides. The sound functioning of the cell depends on good bumper and crumple zone design. Note that, to protect passengers against crushing, it has been proposed in the USA that car roofs must not collapse when they fall from a height of 0.6 m by more than 100 mm; this standard is met by some designs.

Fibre-reinforced plastic car bodies, when involved in collisions, were found to disintegrate and to be especially vulnerable to side impacts. Their design and performance has been greatly improved to be comparable to, and sometimes better than steel body cars; see below.

CRASH CUSHION DESIGN

The *modular crash cushion* which consists of a regular arrangement of tight-head drums, see Figure 2.5, is a vehicle impact attenuation system, so positioned as to protect motorists from driving hard into nominally rigid obstacles; the aim is to arrest cars in 3.6 m to 5.5 m. Full-scale tests showed the proposed arrangements to be effective. Much research work has been centered around the behaviour of tube clusters used in highway applications as modular crash cushions. Portable versions, cantilevered off the back of service vehicles to provide protection during maintenance op-

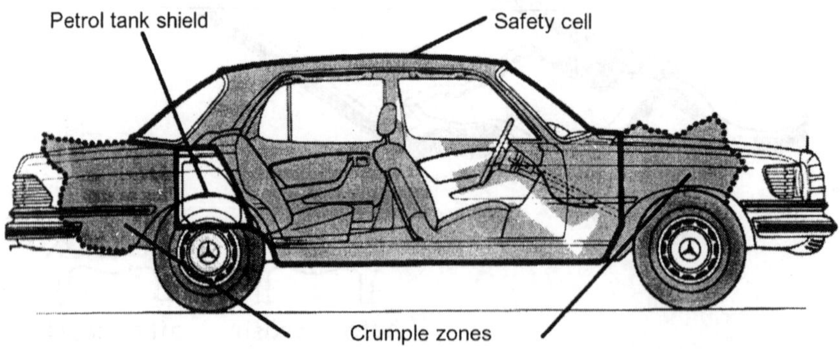

Figure 2.4. Rigid passenger cell with impact absorbing crumple zones.

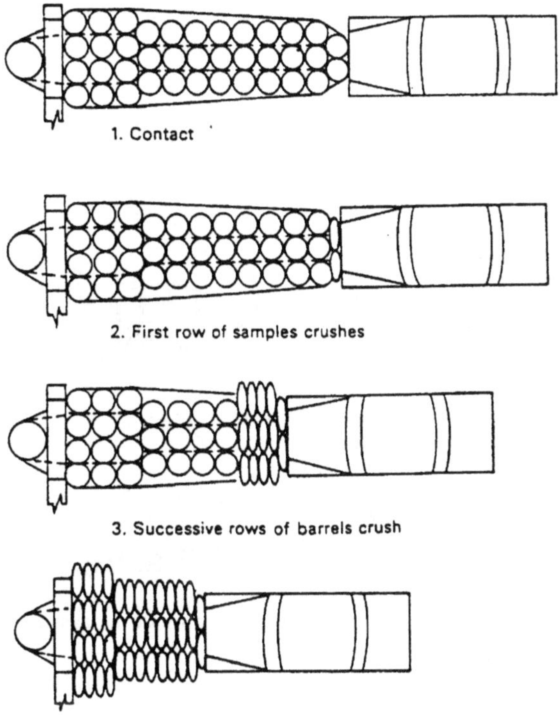

Figure 2.5. Successive crushing of modular crash cushion.

erations, and stationary, triangular arrays, located close to bridge abutments, etc. to trap or redirect errant vehicles, have been considered [1, 7].

AIRFRAME CRASHWORTHINESS

It refers to the capability of an airframe structure to maintain a protective shell around occupants during a crash, and ability to subject any human occupant to retardation rate which can be tolerated. The topic is outlined in Reference [1].The design of crashworthy envelopes must necessarily be coupled with knowledge of the loads, i.e. large retardation or acceleration rates which human occupants can withstand.

Note that, in achieving these, a structure which, on making contact with the earth, reduces the gouging and scooping of soil and, therefore, reduces retardation rates and associated forces, must be designed. Reinforcing cockpit and cabin structures to withstand large forces with little, or without any, plastic deformation and crashworthiness, as it touches on wings, empennage, landing gears and engine mounts of the aircraft, have to be also considered.

The scooping-up and ploughing of earth by a collapsed fuselage, see Figure 2.6, contributes to increased rate of retardation, because a mass of earth must be quickly accelerated in the scooping operation, i.e. there is momentum exchange, and the

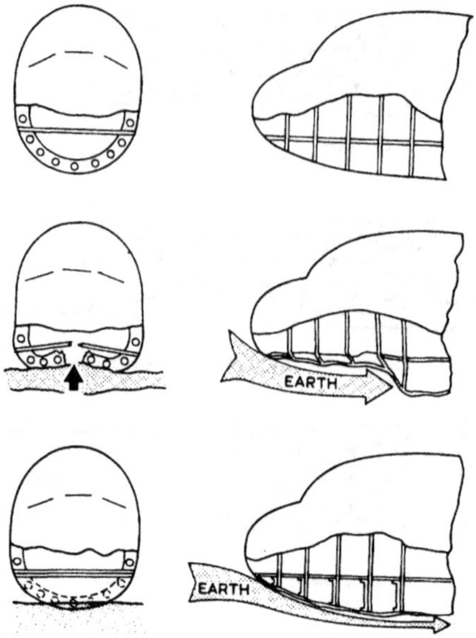

Figure 2.6. Method of reinforcing nose structure to provide increased resistance to vertical loads and to reduce earth scooping.

ploughing results in an increased force being continuously applied during the whole retardation process. The severity of the primary impact process very much depends on the distance available for dissipating any initial kinetic energy. For aircraft impinging against the earth with a velocity vector substantially parallel to it, there is usually a relatively long distance available for arrest in scooping and ploughing, whilst in the case of the velocity vector making a large angle with the earth surface, arrest must be relatively rapid with retardations and forces generally much higher than in the horizontal case. Longitudinal crashworthiness is enhanced by designing the aircraft to reduce scooping and ploughing at the fuselage front and belly, so that a surface skidding process is best aimed for, see Figure 2.6. Besides considering cabin floor structure strength, some compressive buckling failure of the fuselage requires to be envisaged. Vertical impacts can be minimised by providing seats with a relatively long vertical energy absorbing stroke and/or including a sub-cabin floor crushable region of depth sufficient to adequately absorb the vertical energy of impact.

2.2 USE OF COMPOSITE MATERIALS IN CRASHWORTHINESS APPLICATIONS

This field, besides being of fast growing significance to the vehicle manufacturers, in itself merits great attention and interest from the mass of professional engineers and engineering teachers; a review on the topic can be found in Reference [9].

Since late 1982, a team from Ford Engineering and Research Staffs has been developing a variety of vehicle structures made of glass fibre reinforced plastics (FRP) [10]. Projects have included: leaf springs, a rear floor pan, a front rail and apron structure for the Escort and a complex, number one crossmember for an AWD Aerostar. Most test results and driving impressions from vehicles modified to incorporate these structures indicate that potential noise, vibration and harshness (NVH) and/or weight benefits were achievable. However, it was not known, whether the NVH/weight benefits of the hybrid metal/FRP structures, such as the Escort with composite rear floor pan or composite front stucture, would carryover into a complete FRP body structure. As a first step to answering this question, in mid- 1986 it was decided to build and test a complete Taurus FRP vehicle body structure utilizing the resin transfer molding (RTM) process. Shell Chemical Company (later in the form of their subsidiary, Ardyne, Inc.) became a partner with Ford and agreed to create the molds and fabricate the composite parts. The structure was divided into five major assemblies, see Figure 2.7.

The enormously successful use of composites, particularly from the point of view of crashworthiness, in Formula One racing cars is described by O' Rourke [11] and Savage [12]. Whilst it should be realised that cost is a much less prohibitive factor in Formula One than in more conventional vehicle development, this should not mean there will not be a transfer of technology and experience from this area to more cost-sensitive vehicle applications.

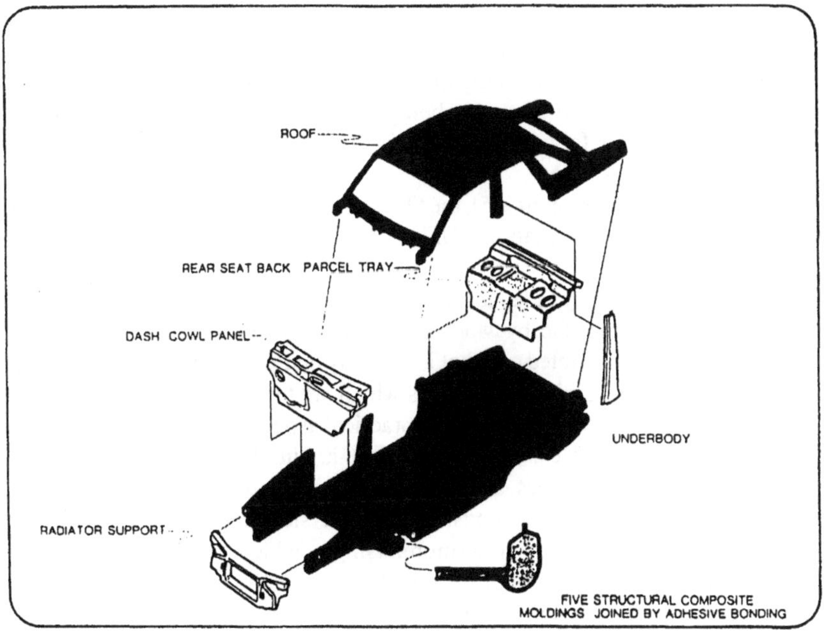

Figure 2.7. Ford Taurus "TUB" concept: all composite body structure.

Composites are used extensively for secondary components on current commercial production aircraft. Committed production of control surfaces, such as ailerons, elevators, rudders and spoilers was initiated in 1982 on the Boeing 767. Corresponding levels of composite applications followed on the 757 (1983), 737-300 (1984) and the 747-400 (1988). These control surfaces and similar fairings, panels, and nacelle components are typically 20–30% lighter than their conventional metal designs. Generally fabricated as a sandwich, their face sheets are carbon fibre or hybrid with carbon, aramid or fibreglass [13].

The V-22 Osprey tilt rotor aircraft, developed by Bell Helicopter Textron and Boeing Helicopters, incorporates over five tonnes of composite materials in the static and dynamic airframe structure. Frames were made of a hybrid of AS-4 woven fabric and IM-6 unidirectional tape, both with 3501-6 epoxy resin. Honeycomb sandwich structure is also used extensively on the V-22 in secondary structural areas, where easy replacement is possible should need arise. Sandwich panels have a high bending stiffness over weight ratio and are therefore an excellent choice for applications involving out-of-plane loads. Examples are fuel sponsons, floors, landing gear doors, cabin and cockpit doors, fairings, and access panels. Sandwich construction results in a low part count as a result of eliminating stiffeners, and is, therefore, attractive from a cost point of view [14], but see also Reference [6].

British Rail Research conducted quasi-static laboratory crush tests on a full-size crashworthy railway cab structure. Included in the cab design were replaceable energy absorption devices consisting of two glass-reinforced plastic (GRP) tubes, similar to those described above, one mounted at each buffer position. The tubes were designed to collapse axially at a constant combined load of 800 kN, but as Scholes and Lewis [15] reported, in the tests crushing actually commenced at 5% above this value. Furthermore, the subsequent failure of the tubes did not meet the specified requirements, with the absorbed energy being less than half that required. This was attributed to gross shear failure of the matrix, causing the tubes to break into large pieces rather than delaminate.

As referred in the same report [15], the Office for Research and Experiments (ORE), the research arm of the International Union of Railways (UIC), conducted full-scale dynamic collision tests. These tests also included GRP tubes, mounted in the buffer positions, although again they failed to work properly. They also actually failed at a force in excess of their design value, by which time other elements of the cab structure had already collapsed. It was pointed out, that it is clearly undesirable for the cab structure to fail before the sacrificial energy absorbers which are there to protect it.

Lin and Mase [16] discuss some of the practical aspects of introducing energy absorption devices to vehicles. They conclude that simple devices, such as honeycombs or axially crushed tubes, are more effective as add-on energy absorption devices than more complicated methods, such as inversion tubes and the foam filling of primary structural members. Furthermore, using a vehicle impact simulation, they show that it is theoretically possible to absorb extra crush energy, without increasing the crush distance and passenger car deformation, by inserting such energy absorption devices in unutilised space along existing load paths. However, they do acknowledge that other design aspects, such as interference with vehicle function, vehicle servicing and, especially, directionality of the devices must be taken into account.

Whilst crash energy management is one of the primary design requirements that the front-end structure of a vehicle must meet, other factors also exist, which should be taken into account, when seriously considering the large scale use of composite materials. Tong [17] and Johnson [18] both highlight the importance of fire hazard and toxication in the event of a serious crash and Noton [19] lists the necessity for fire-retardant additives as a potential disadvantage of the use of glass-reinforced plastics for rail transportation applications. However, railway rolling stock's increasingly widespread use of glass fibre reinforced phenolic mouldings, which possess excellent fire resistance, smoke emission and toxicity characteristics, is reported in Reference [20]. Furthermore, the potential of lightweight rail vehicle body shell panels made of a carbon fibre reinforced plastic with a phenol resin is discussed by Suzuki and Satoh [21]. This material, certified as non-flammable, was shown to outperform aluminium at high temperatures in bending stress-deformation tests.

As far as enviromental damage due to the effect of high temperature is concerned, Thornton and Jeryan [22] note that a reduction in performance can also arise from water absorption. The combination of high temperature and high humidity is known to be particularly detrimental in reducing tensile stress and fatigue strength, although little information appears to be known about their effect on energy absorption capability.

Finally, the issue of cost should not be avoided. However, the fabrication of composite allows for many variants and so it is extremely difficult to obtain easily-comparable, up-to-date information. On a basic material property level, Reference [23] gives approximately order of magnitude cost data for a wide range of material types. The relevant information has been extracted and is presented in Table 2.1, but cost data for aramid-epoxy composites are not included. However, Ashby [24], including material property data presented in the same way as Reference [23], places its cost somewhere between that of carbon-epoxy and glass-epoxy composites. An estimated cost of 10–70 USD/kg may not be unreasonable. However, the following point must be outlined when the parameter cost is considered. There is a fundamental difference in the strategy of application of composites between the aerospace industry and the automobile industry; this is primarily due to the volume requirements of the two businesses. In aerospace and defense, the design of the structure is optimised to provide the required functionality and performance and the manufacturing process (and the associated cost) is subsequently selected on the basis that the process is capable of achieving the desired design. In direct contrast, in high-volume production industries such as the automotive industry, the rate of manufacture is critical to satisfying the economics of this consumer industry. Thus, manufacturing processes, which are capable of satisfying production output, are the primary consideration and design of a component or structure must be within the boundary constraints of the selected fabrication process.

However, care should be exercised when comparing the costs of different materials for a certain application. For example, although the cost in USD/kg of aluminium and composites exceeds that of the carbon steels, the lower density of these materials means that less mass may be needed in a given case. Furthermore, these figures are given for the material in an unfinished state. Additional processing will add to the price of the final product and, indeed, McCarty [25] suggests (in the context of the commercial aircraft industry) that the difference between "in the door" and "out the door" costs of composites in a given application may well be considerably less than those of aluminium. As an alternative comparison, Frame [26] estimates that aircraft specification carbon composites cost about 15 times more than conventional aluminium alloys.

CHAPTER 3

FAILURE MECHANISMS OF COMPOSITES

3.1 NOTATION

b = width
E_1 = longitudinal modulus of elasticity
E_2 = transverse modulus of elasticity
F = force
G = strain energy release per unit width
G_{12} = laminate shear modulus
M = bending moment
R_{ad} = fracture energy
S = laminate shear strength
V_f = fibre volume fraction
X_c = ultimate longitudinal compressive strength
X_t = ultimate longitudinal tensile strength
Y_c = ultimate transverse compressive strength
Y_t = ultimate transverse tensile strength
γ_{12} = laminate shear strain
γ_{12}^u = ultimate shear strain
ε_1 = longitudinal strain
ε_2 = transverse strain
ε_1^c = ultimate longitudinal compressive strain
ε_1^t = ultimate longitudinal tensile strain
ε_2^c = ultimate transverse compressive strain
ε_2^t = ultimate transverse tensile strain
ν_{12} = major Poisson ratio
σ_1 = longitudinal stress
σ_2 = transverse stress
τ_{12} = laminate shear stress

3.2 FAILURE MODES

In a very broad sense, failure of a structural element can be stated to have taken place when it ceases to perform satisfactorily. Therefore, the definition of failure may change from one application to another. In some applications a very small deformation may be considered failure, whereas in others only total fracture or separation constitutes failure. In the case of composite materials, internal material failure generally is initiated long before any change in its macroscopic appearance or behaviour is observed. The internal material failure may be observed in many forms, separately or jointly, such as (a) breaking of the fibres, (b) microcracking of the matrix, (c) separation of fibres from the matrix in the form of (i) debonding or/and (ii) pull-out and (d) separation of laminae from each other in a laminated composite (called delamination). The effect of internal damage on macroscopic material response is observed only when the frequency of internal damage is sufficiently high. Agarwal and Broutman [27] report a detailed description of the various failure modes of fibre-reinforced composite materials and, moreover, of the main characteristic failures occurred when an orthotropic lamina is subjected to various loading conditions.

FIBRE BREAKAGE

Whenever a crack has to propagate in the direction normal to the fibres, fibre breakage will eventually occur for complete separation of the laminate. Fibres will fracture when their fracture strain is reached. Although the fibres are responsible for imparting high strength to the composites, the fracture of fibres accounts for only a very small fraction of the total energy absorbed.

MATRIX DEFORMATION AND CRACKING

The matrix material surrounding the fibres has to fracture to complete the fracture of the composite. Thermosetting resins, such as epoxies and polyesters, are brittle materials and can undergo only a limited deformation prior to fracture, whilst metal matrices may undergo extensive plastic deformation. The energy absorbing mechanism of polymer matrices is characterised by cracking and small deformation and, therefore, their contribution to the total energy absorbed is relatively insignificant as compared to the metal matrix composites.

FIBRE DEBONDING

During the fracture process the fibres may separate from the matrix material by cracks running parallel to the fibres (debonding cracks). In this process, the chemical or secondary bonds between the fibres and the matrix material are broken. This type of cracking occurs when fibres are strong and the interface is weak. A debonding crack may run at the fibre-matrix interface or in the adjacent matrix, depending on their relative strengths. With decrease in interface strength extensive debonding may be obtained owing, therefore, to a significant increase in energy absorbed.

FIBRE PULL-OUT

Fibre pull-outs occur when brittle or discontinuous fibres are embedded in a tough matrix. The fibres fracture at their weak cross sections that do not necessarily lie in the plane of composite fracture. The stress concentration in the matrix produced by the fibre breaks is relieved by matrix yielding, preventing, therefore, a matrix crack that may join the fibre fractures at different points. In such a case, the fracture may proceed by the broken fibres, being pulled-out of the matrix, rather than fibres fracturing again at the plane of composite fracture. This is particularly applicable to those fibres whose ends lie within a small distance (less than half the critical fibre length) of the particular cross section at which failure occurs.

Note that, fibre debonding and fibre pull-out may appear to be similar phenomena, because of failure taking place at the fibre matrix interface in both cases, however, fibre debonding takes place when a matrix crack is unable to propagate across a fibre, whereas fibre pull-outs are a result of the inability of a crack, initiated at a fibre break, to propagate into the tough matrix. The fibre pull-outs are usually accompanied by extensive matrix deformation, which is absent in fibre debonding.

DELAMINATION CRACKS

A crack propagating through a ply in a laminate may be arrested as the crack tip reaches the fibres in the adjacent ply. This process of crack arrest is similar to the arrest of a matrix crack at the fibre matrix interface. Due to high shear stresses in the matrix adjacent to the crack tip, the crack may be bifurcated and start running at the interface parallel to the plane of the plies. These cracks are called delamination cracks and, whenever present, they are responsible for absorbing a significant amount of fracture energy. Delamination cracks frequently occur when laminates are subjected to flexular loading, as is the case of the Charpy and Izod impact tests.

3.3 FAILURE UNDER SEVERAL LOADING CONDITIONS

3.3.1 Failure Under Tensile Loading

LONGITUDINAL TO THE FIBRES

In a unidirectional composite, consisting of brittle fibres subjected to increasing longitudinal tensile loading, failure initiates by fibre breakage at their weakest cross sections. As the load increases, more fibres break. Breaking of the fibres is a completely random process. As the number of broken fibers increases, some cross section of the composite may become too weak to support an increased load, thus, causing a complete rupture of the composite. The interfaces of the broken fibres may become debonded, because of stress concentrations created at the fibre ends and, thus, they may contribute to the separation of the composite at a given cross section. In other cases, cracks at different cross sections of the composite may join-up by debonding of the fibres along their length or by shear failure of the matrix. Therefore, a unidirectional composite can fail in at least three modes under longitudinal tensile load. These models are: (1) brittle, (2) brittle with fibre pull-out, (3) brittle failure with fibre pull-out and (a) interface-matrix shear failure or (b) constituents debonding, i.e. matrix breaking away from the fibres [28].

TRANSVERSE TO THE FIBRES

Fibres perpendicular to the loading direction act essentially to produce stress concentrations at the interface and in the matrix. Therefore, unidirectional composites, subjected to transverse tensile loading, fail because of matrix or interface tensile failure, although in some cases they may fail by fibre transverse tensile failure, if the fibres are highly oriented and weak in the transverse direction. Thus, the composite failure modes under transverse tensile load may be described as: (1) matrix tensile failure and (2) constituent debonding and/or fibre splitting. Matrix tensile failure with constituent debonding means that some portions of the fracture surface are formed, because of failure of interfacial bonds between the fibres and the matrix [29].

3.3.2 Failure Under Compressive Loading

LONGITUDINAL TO THE FIBRES

When composites are subjected to compressive loading, continuous fibres act as long columns and microbuckling of the fibres can occur. In composites with very low fibre volume content, fibre microbuckling may occur even when the matrix stresses are in the elastic range. However, at practical fibre volume fractions ($V_f > 0.40$), fibre microbuckling is generally preceded by matrix yield and/or constituent debonding and matrix microcracking. Compressive failure of a unidirectional composite, loaded in the fibre direction, may be initiated by transverse splitting or failure of the composite, i.e. the transverse tensile strain, resulting from the Poisson ratio effect, can exceed the ultimate transverse strain capability of the composite resulting in cracks at the interface [30, 31]. Shear failure constitutes another mode of gross failure. Subsequently, the failure modes for composites, subjected to longitudinal compressive loading, may be listed as: (1) transverse tensile failure, (2) fibre microbuckling (a) with matrix still elastic, (b) preceded by matrix yielding, (c) preceded by constituents debonding, and (3) shear failure.

TRANSVERSE TO THE FIBRES

A unidirectional composite subjected to transverse compressive loading, generally fails by shear failure of the matrix, which may be accompanied by constituent debonding or fibre cracking. Therefore, the failure modes of a unidirectional composite under transverse compressive loading may be listed as: (1) matrix shear failure, (2) matrix shear failure with constituent debonding and/or fibre cracking. Experimental investigations of Coolings [32] indicate that, when the composites are subjected to transverse compressive loading, failure occurs by shear in a direction normal to the fibres, on planes parallel to them at the expected angles. It has been suggested, that the failure is precipitated by failure of the fibre-resin bond, thus, the transverse compressive strength is lower than the longitudinal compressive one. However, if constraints are placed on the specimen to prevent its deformation in the direction perpendicular to the plane of load-fibre axes, it is possible to achieve a transverse compressive strength comparable with the longitudinal compressive one. The increase in strength is observed because failure now occurs by the shear failure of the fibres, whose strength is higher than the matrix strength or the bond strength. In this mode of failure, that is the shear failure of the fibres, it has been observed, as expected, that the transverse compressive strength of the composite increases with increasing fibre volume fraction.

3.3.3 Failure Under In-Plane Shear Loads

In this case, the failure could take place by matrix shear failure, constituent debonding, or a combination of the two. Thus, the failure modes are: (1) matrix shear failure, (2) matrix shear failure with constituent debonding and (3) constituent debonding.

3.3.4 Failure Under Bending

The micromechanism of failure of a laminate subjected to bending is characterised by the development of two distinct regions, i.e. a top compressive zone and a bottom tensile zone, with different microscopic characteristics. The main features of the regions under compressive loading can be classified as: (i) slipping and/or fracturing of the matrix material; (ii) delamination cracks, which propagate longitudinally through the wall thickness; (iii) small cracks, which develop perpendicular to the longitudinal axis of symmetry. In the bottom tensile region, the associated failure mechanism mainly results in: (i) a "brush like" appearance of pulled-out fibres and/or matrix cracking; (ii) fibre/matrix interface debonding of transverse fibres; (iii) delamination and fibre breakage of longitudinal ones [33].

3.3.5 Failure Under Impact Loading

The load-time history of an impact loaded composite material can be divided into two distinct regions, a region of fracture initiation and a region of fracture propagation. As the load increases during the fracture initiation phase, the elastic strain energy is accumulated in the specimen without gross failure. Note that microscopic failure mechanisms, such as microbuckling of the fibres on the compression side or debonding at the fibre-matrix interface, may occur. When the critical load is reached at the end of the initiation phase, the composite specimen may fail either by a tensile or a shear failure, depending on the relative values of the tensile and interlaminar shear strengths. At this point, fracture propagates either in a catastrophic "brittle" manner or in a progressive manner, continuing to absorb energy at smaller loads [27].

3.4 MATERIAL TESTING

For the analysis and, finally, the design of a structure, which subjected to loading, it is necessary to primarily perform the experimental characterisation of the material used. Experimental characterisation refers to the determination of the material properties through tests conducted on suitably designed specimens. Understanding the material response over the entire range of loading is necessary, if advanced design procedures are employed for efficient material utilisation. In the case of composite materials, it may be desirable to begin the design with constituent material properties and to arrive at the composite macro-mechanical properties through micro-mechanics analyses.

Elastic constants and strengths constitute basic mechanical properties of materials. In the case of a laminate, the interlaminar shear strength is also an important property. It is necessary to establish these properties for the minimum characterisation of a unidirectional lamina. However, since composite material structures are often subjected to bending, it may be necessary to establish flexural properties in addition to the properties just mentioned. All properties may be associated to a single ply or lamina of the composite material, that is the basic building block for composite structures, and, therefore, the laminate theory can be used to calculate the properties of laminates. However, practical considerations often prevent the construction of single-layer test specimens and, therefore, it becomes necessary to conduct tests on multilayer specimens and use the appropriate lamination theory in terms of lamina properties. If the laminates are unidirectional, their behaviour obviously simulates the lamina behaviour. A brief description of the more significant experimental tests is given below; for more details, see Agarwal and Broutman [27] and Carlson and Pipes [34].

3.4.1 Uniaxial Tension Test

The static uniaxial tension test is probably the simplest and most widely used mechanical test. This test is conducted to determine elastic modulus, tensile strength and Poisson ratio of the material. In the case of composite materials, the tension test is generally performed on flat specimens. The most commonly used specimen geometries are the dog-bone specimen and the straight-sided specimen with end tabs. A uniaxial load is applied through the ends by providing a pin-type or serrated jaw type and connection. The dog-bone specimens may fail at the neck radius, particularly when testing uniaxial specimens, because of stress concentrations and poor axial shear properties of the specimen. The pin-type and connections also tend to fail at low loads by shear.

The ASTM standard test method for tensile properties of fibre resin composites has the designation D3039-76 (reapproved 1982). It recommends that the specimens with fibres parallel to the loading direction must be 12.7 mm wide and made of 6–8 plies, and the specimens with fibres perpendicular to the load must be 25.4 mm wide with 8–16 plies. Length of the test section in both cases should be 153 mm.

The data recording in a tension test consist of measuring the applied load and strain, both parallel and perpendicular to the load. The applied load is usually measured by means of a load cell that is generally provided with the testing machine. The strains can be measured by means of an extensometer or an electrical-resistance strain-gage. Extensometers, mechanically attached, tend to slip at times, although they are quite simple to use. Strain-gage may be used for a more accurate measurement of strains. From these data taken until failure, stress-strain curves can easily be plotted for the material and the required material properties determined. If the load is applied in the longitudinal direction, the initial slope of the stress/strain curve provides the longitudinal modulus of elasticity, E_1. Similarly, the transverse modulus of elasticity, E_2 can be determined by applying the load in the transverse direction. The ultimate longitudinal and transverse tensile strengths, X_t and Y_t, respectively, are obtained from the knowledge of load at fracture in the two tests. The Poisson ratio, ν_{12}, is obtained from the strains parallel and perpendicular to the load, measured at the same axial load.

3.4.2 Uniaxial Compression Test

Static uniaxial compression tests are similar to the tension ones, but they are more problematic. The biggest problem is the necessity to prevent geometric buckling of the specimens. This requirement is particularly relevant to thin, flat specimens and is usually met by providing multiple side supports that prevent the specimen from buckling out of its plane. The use of side supports can be avoided by using a block or bar-type specimen rather than a plate. However, the block-type specimens are more difficult to be prepared.

Witney et al. [35] report on a number of loading fixtures and specimen configurations, that have been developed to measure the compressive strengths of composites. Alignment of the test fixture and specimen constitutes an essential consideration in any compression test. The most common test fixture is the IITRI test fixture, developed by the Illinois Institute of Technology Research Institute and reported by Hofer and Rao [36]. A relatively short (gage length 12.7 mm), unsupported test specimen is used. The fixture employs linear borings and hardened steel shafts to ensure co-linearity of the loading direction. The data recording in compression tests is also similar to that in tension tests. The strain in the direction of loading may be measured by a compressometer or electrical-resistance strain-gages. The testing machine head movement is not a reliable measure of strain, because some error may result from crushing of the specimen ends. The strain in the direction perpendicular to the load is measured using strain-gages only because of the space limitations. From these measurements the elastic moduli and Poisson ratio of the material in compression can be determined and the stress/strain curve in compression plotted. A straight-sided specimen is well suited for elastic moduli and Poisson ratio determination. For compressive strength determination, reduced-center-section specimens are often used to ensure that failure does not occur near the end of the specimen. It should be mentioned, that the composite properties in compression may also be determined by using sandwich-beam specimens or filament wound tubes or ring specimens. However, these types of specimen have not been extensively used as the flat specimens.

3.4.3 In-Plane Shear Test

The tests, in which shear distortion takes place entirely in the plane of the composite material sheet, are termed as in-plane shear tests. The properties that are determined through these tests are the shear modulus and the shear strength. In these tests, a material specimen is subjected to loading that produces a pure shear state of stress and the resulting strains are measured. The easiest way to produce a state of pure shear is to subject a thin-walled circular tube to a torque about its axis. Due to the difficulties associated with the fabrication and testing of tubular specimens, there are other test methods that are employed for the determination of the in-plane shear properties of unidirectional composites. Among them, the ±45 laminate test, and the off-axis tensile test are currently the most commonly employed methods.

$[\pm 45]_S$ COUPON TEST

In-plane shear properties of a unidirectional composite can be determined by conducting a tension test on a $[\pm 45]_S$ laminate. The test results are interpreted using laminate stress analysis. These results show that, for the $[\pm 45]_S$ laminate, the laminate shear stress, τ_{12} and the corresponding shear strain, γ_{12} can be obtained from the laminate stress and strains. Thus, the results of a tension test on a $[\pm 45]_S$ laminate can be employed to establish shear stress/strain response of the lamina. This test method is quite suitable for determination of shear modulus, G_{LT}, but not for shear strength, S, because the lamina is in a state of combined stress rather that pure shear. This test procedure was suggested by Rosen [37], and details are presented in the ASTM standard 3518-76 (reapproved 1982).

OFF-AXIS COUPON TEST

The off-axis tension test is often used to determine the in-plane shear response of unidirectional composites. Pindera and Herakovich [38] have suggested the use of 45° off-axis specimen for the measurement of G_{12}. In the case of low off-axis configurations, such as the 10° off-axis coupon, a specimen with very high aspect ratio must be employed. The 45° off-axis specimen is not suitable for measurement of the shear strength due to the presence of σ_1 and σ_2. Chamis and Sinclair [39] recommended the use of the 10° off-axis test to minimize the effects of σ_1 and σ_2 on the shear response.

3.4.4 Uniaxial Bending (Flexural) Tests

The two most popular flexural tests are the three-point and four-point bending tests. In these tests, a flat specimen is simply supported at the two ends and it is loaded by either a central load (three-point bending) or by two symmetrically placed loads (four-point bending). Flexural response of the beam is obtained by measuring the applied load and the corresponding strain. The strain may be measured using a strain-gage bonded to the beam on its tension side, i.e. on surface opposite to the applied load. The bending moment, M is determined from the measured load and specimen geometry, and the related stress is then calculated; therefore, the complete stress/strain behaviour in bending may be obtained. Note that the flexural strength is the theoretical value of the stress at failure on the surface of the specimen. It is calculated from the maximum bending moment by assuming a straight-line stress/strain relation to failure. Relevant standard for flexure test methods is the ASTM standard D790-1. Some useful information on flexure test methods is also given in Reference [40].

As observed in the tests of axial splitting crush, the laminate can be bent over a small radius without fibre breakage, due to delamination, which occurs simultaneously with bending failure. In ordinary three point bending tests, the same material could not be bent over an equally small radius before fracture. In order to evaluate the energy absorbed in the sharp bending process by an experimental approach, the failure mode must be somehow reproduced. Yuan and Viegelahn [41] suggest an experimental test, designated as curling test, for the purpose of reproducing the failure mode observed in the axial splitting crush. Curling tests are conducted by pushing 25.4 mm wide composite specimens through the curling test fixture. Pushing through the radiused corner of the test fixture, delamination occurs and the delaminated plies are bent over the small radius without figure breakage. The forces required to push specimens through the curling test fixture are measured and, therefore, the energy absorbed due to bending for a unit width of material may be estimated.

3.4.5 Determination of the Interlaminar Shear Strength and Fracture Toughness

The stresses acting at the interface between two adjacent laminae are called interlaminar stresses; these stresses cause relative deformations between the two lamina. If they are sufficiently high, they may cause failure along the interface. It is, therefore, of considerable interest to evaluate interlaminar shear strength through tests in which failure of laminates initiates in a shear (delamination) mode; see References [42, 43].

A flexural test with a short span called short-beam shear test, is the most widely used test to evaluate interlaminar shear strength. As mentioned above, the maximum shear stress in a beam occurs at its mid-plane. Therefore, in a short-beam shear test, failure consists of a crack running along the mid-plane of the beam so that the crack plane is parallel to the lamination plane. It may be emphasized, that a short-beam shear test becomes invalid, if the tensile failure of fibres precedes the shear failure or, if tensile failure and shear failure occur simultaneously. Another type of test used for evaluating shear strength (sometimes called thickness shear) is the notched-plate test. In this test, two notches are provided in the thickness direction from the opposite faces of the specimen and the specimen is tested in tension or compression. The distance between the notches is adjusted such that the failure load, corresponding to interlaminar shear failure between the notches, is smaller that the failure load corresponding to tensile failure of the notched cross sections. However, the short beam shear test is much simpler to perform than the notched plate test and, hence, is very widely used.

Because of the importance of the delamination failure mode in laminated composite structures, the static interlaminar fracture toughness of unidirectional composites is often determined. The double cantilever beam (DCB) test is commonly used for this purpose. This test was initially developed for investigations of adhesive fracture mechanics [44]. Because of the similarity between the debonding process in metals and the delamination process in composites, Bascom et. al. [45] and Wilkins et. al. [43] developed the utilized DCB specimens for composites. The specimen is made of an even number of layers. A starter crack at one end of the specimen is introduced at the mid-thickness during fabrication. At the crack end of the specimen, two hinges are mounted for load application. The load/displacement curve during loading is recorded on a chart. The displacements are measured from an extensometer or linear voltage differential transformer (LVDT) attached to the specimen. The strain energy release per unit width of the specimen, G can be obtained using the results of the load/displacement curve and the geometrical characteristics of the specimen tested.

Kendall [46], dealing with the interfacial cracking of composite materials, suggests a simple experimental method to determine the fracture energy, R_{ad}, which is required to fracture a unit area of the adhesive at the interface between two adjacent layers. This test, designated as peel test, appears some similarities with DCB test, in the form of specimens used. The long arms attached to the model composites are pulled apart on an Instron machine with a force F_{peel}. Therefore, the adhesive fracture energy could then be calculated from the relationship $R_{ad} = 2\,F_{peel}/b$, where b is the specimen width.

3.4.6 Impact

High strain-rate or impact loads may be expected in many of the engineering applications of composite materials. The suitability of a composite for such applications is, therefore, determined not only by the usual design parameters, but by its impact or energy absorbing properties. Frequently, an attempt to improve the tensile properties results in the determination of impact properties. For example, the high-modulus fibre composites are more brittle than the lower-modulus fibreglass composite materials. Thus, it is important to have a good understanding of the impact behaviour of composites for both safe and efficient design of structures and to develop new composites having good impact properties as well as good tensile ones. A very common way to evaluate impact properties is to determine material toughness by measuring the energy required to break a specimen of a particular geometry. The well-known Charpy and Izod impact tests, developed for isotropic materials, are widely used for this purpose.

3.5 FAILURE CRITERIA

A design engineer estimates the load-carrying capacity of a structure or a component through the procedures that involve stress analysis of the structure and a comparison of the actual stress field with the strengths of the material. When the actual stress field is simple, such as the one produced in the specimens during strength determination tests, a direct comparison can be made regarding the load-carrying capacity of the structure obtained, otherwise, a direct comparison may not be valid. In reality, structures may be subjected to biaxial stress states, and it is impractical to establish the strength characteristics of materials for every possible biaxial stress state. A "failure criterion" or "failure theory", if valid, can predict the strength of materials under biaxial stress states using strength data obtained from uniaxial tests. For isotropic materials, the failure criteria are written in terms of principal stresses and ultimate tensile, compressive, and shear strengths. In the case of orthotropic materials, the situation is considerably more complex. The most important complexity arises from the fact that their strengths, like their elastic constants, are direction dependent. Thus, for an orthotropic material, an infinite number of strength values can be obtained, even through uniaxial tests, depending on the direction of load application. For prediction purposes, they can be limited to five strengths in the principal material directions, see Section 3.4 above. It may be pointed out here, that a uniaxial stress, applied in any direction other than the principal material axes, produces multiaxial stresses along the principal material axes. Therefore, off-axis uniaxial strengths of orthotropic materials, like their strengths under complex stress states, must be predicted through an appropriate failure criterion. All failure criteria for orthotropic materials are, quite obviously, expressed in terms of stress along principal material axes rather than the principal stresses.

There are a number of theories for prediction of the failure of isotropic and orthotropic materials subjected to a complex stress state. Nahas [47] reports on a useful survey of failure and postfailure theories of laminated fibre reinforced composites listing also a large number of references. Many failure theories for orthotropic materials have been devoted to the failure theories of isotropic materials. Some of them are only applicable to specific types of composites. Four of the failure criteria widely used for fibre-reinforced composite materials are discussed here, see also References [48, 49] for details. Note that these criteria cannot predict the failure mode but only the occurence of failure.

Consider a unidirectional lamina, see Figure 3.1, with the reference axes coinciding with the axes of symmetry; designated as the longitudinal and transverse directions. For a unidirectional lamina or composite, there are four independent elastic constants, namely, the elastic moduli in the longitudinal and transverse directions, E_1 and E_2, respectively, the shear modulus or the modulus of rigidity associated with the axes of symmetry, G_{12}, and the major Poisson ratio, ν_{12}, providing the transverse strain caused by a longitudinal stress. Moreover, there are five independent strengths, namely, the tensile (X_t, Y_t) and the compressive (X_c, Y_c) strengths in the longitudinal and transverse directions, respectively, and the in-plane shear strength, S.

3.5.1 Maximum Stress Theory

Hill [50] states that failure will occur if any of the stresses in the principal material axes exceeds the corresponding allowable stress. Thus, to avoid failure, the following inequalities must be satisfied: for the tensile stresses

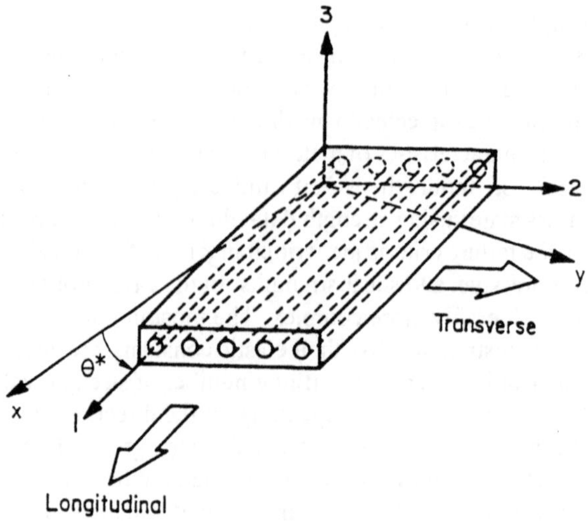

Figure 3.1. Schematic diagram of a unidirectional lamina.

$$\begin{aligned} \sigma_1 &< X_t \\ \sigma_2 &< Y_t \\ |\tau_{12}| &< S \end{aligned} \tag{3.1}$$

and for the compressive stresses

$$\begin{aligned} \sigma_1 &> X_c \\ \sigma_2 &> Y_c \end{aligned} \tag{3.2}$$

According to this theory, when one of the inequalities indicated in Equations (3.1) and (3.2) is violated, the material is considered to have failed by a failure mode associated with the allowed stress. There is no interaction between the modes of failure in this criterion. Therefore, this actually constitutes not one criterion but five sub-criteria.

3.5.2 Maximum Strain Theory

This theory [50] states that failure may occur if any of the strains in the principal material axes exceeds the corresponding allowed strain. Thus, the following inequalities must be satisfied for the "no failure" mode,

$$\begin{aligned} \varepsilon_1 &< \varepsilon_1^t \\ \varepsilon_2 &< \varepsilon_2^t \\ |\gamma_{12}| &< \gamma_{12}^u \end{aligned} \tag{3.3}$$

If the normal strains are compressive

$$\varepsilon_1 > \varepsilon_1^c$$
$$\varepsilon_2 > \varepsilon_2^c \tag{3.4}$$

The maximum strain theory is similar to the maximum stress theory. All stresses are replaced by the corresponding strains, first to apply to the maximum strain theory. If the material is assumed to be linearly elastic, up to the ultimate failure, the ultimate strains (allowed strains) in Equations (3.3) and (3.4) can be related directly to the strengths.

$$\begin{aligned} \varepsilon_1^t &= X_t / E_1 \\ \varepsilon_2^t &= Y_t / E_2 \\ \gamma_{12}^u &= S / G_{12} \\ \varepsilon_1^c &= X_c / E_1 \\ \varepsilon_2^c &= Y_c / E_2 \end{aligned} \tag{3.5}$$

Predictions of the two theories are quite close to each other, since the material has been assumed to be linearly elastic, up to the ultimate failure. The differences are associated to the effect of the Poisson ratio. When the material does not remain linearly elastic up to failure, the two theories are completely independent and have to be applied separately. In that case, larger differences in their predictions should be expected.

3.5.3 Maximum Work Theory

This theory indicates that, under plane stress conditions, failure initiates when the following inequality is violated:

$$(\sigma_1 / X_t)^2 - (\sigma_2 / X_t)(\sigma_2 / X_t) + (\sigma_2 / Y_t)^2 + (\tau_{12} / S) < 1 \tag{3.6}$$

When the normal stresses are compressive, the corresponding compressive strengths are to be used in Equation (3.6). The theory was derived in this form by Tsai [51] from the yield criterion for anisotropic materials, originally proposed by Hill [50]. Therefore, it is sometimes referred to as the Tsai-Hill theory.

3.5.4 Tsai-Wu Tensor Theory

Tsai and Wu [52] suggest a quadratic failure criterion, which is widely known as the Tsai-Wu criterion. For an orthotropic lamina under plane stress conditions, it is

$$F_1\sigma_1 + F_2\sigma_2 + F_6\sigma_6 + F_{11}\sigma_1^2 + F_{22}\sigma_2^2 + F_{66}\sigma_6^2 + 2F_{12}\sigma_1\sigma_2 = 1 \quad (3.7)$$

where, failure under combined stress is assumed to occur when the left-hand side of Equation (3.7) is equal or greater than one. All the parameters of the Tsai-Wu criterion can be expressed in terms of the intrinsic strengths, except F_{12}, where,

$$\begin{aligned} F_{11} &= 1/X_t X_c \\ F_{22} &= 1/Y_t Y_c \\ F_1 &= 1/X_t - 1/X_c \\ F_2 &= 1/Y_t - 1/Y_c \\ F_{66} &= 1/S^2 \\ F_6 &= 0 \\ \sigma_6 &= \tau_{12} \end{aligned} \quad (3.8)$$

F_{12} is a strength interaction parameter which has to be determined from biaxial testing. However, Tsai and Hahn [53] suggest that F_{12} may be estimated as

$$F_{12} = -1/(2X_t Y_t X_c Y_c) \quad (3.9)$$

3.6 NUMERICAL SIMULATION

3.6.1 Material Models

The bi-phase rheological model is a numerical material model for finite-element (FE) analyses, adapted to unidirectional long-fibre reinforced composites or composite fabrics. The stiffness and the resistance of its elements are calculated by superimposing the effects of an orthotropic material phase (matrix minus fibres) and of a unidirectional material phase (fibres), with (or without) deformation compatibility. Each phase (fibres, matrix) is assigned a different rheological law: elastic-plastic/brittle orthotropic for the matrix phase and unidirectional elastic-brittle for the fibres. Upon incremental loading, the stresses are calculated separately in each phase, and damage (matrix cracking, matrix slipping, fibre rupture) can propagate independently, based on the criteria chosen for each phase. A multidirectional laminate is modelled by stacking through the thickness several such elements with the fibres oriented in different directions with respect to a global reference frame [54].

Subsequently, the material model of the matrix phase has been augmented by an elastoplastic (von Mises) material law and by a modulus damage-fracturing material law. This law is also available for the fibre phase.

3.6.2 Analysis Tools

PAM-FISS PROGRAM

The PAM-FISS is a non-linear implicit finite element (FE) quasi-static analysis program, which contains the unidirectional solid element bi-phase material model [55]. This program also contains fracture mechanics analysis techniques, such as an automatic crack advance scheme and several strain energy release rate calculation schemes (G-values). The fracture mechanics techniques have been applied successfully to matrix (resin) intralaminar (splitting and transverse) and interlaminar (edge delamination and outer ply blistering) crack advance [56–58]. A more involved damage mechanics technique, the original (D_c, r_c) critical damage over a characteristic distance material fracture criterion, has been applied successfully to the simulation of the advance of fibre cracks in the critical plies of multilayered (ML) tensile test pieces, see References [59, 60].

A matrix (resin) fracturing modulus damage and strain-softening material model, as well as a plasticity model for matrix damage, have been used to evaluate subcritical matrix damage in tensile test pieces [61]. These laws have also been used for matrix compression damage, with the associated fibre de-confinement and fibre buckling under compressive stress, in composite fabric coupon compression and composite fabric coupon bending test simulations.

The PAM-FISS tool, in conjunction with the bi-phase (fibres, matrix) material description, is particularly well suited for quasi-static tensile, bending, and compression, destructive and non-destructive coupon test simulations on the composite wall micro-scale, from which equivalent models on the wall macro-scale can be derived.

PAM-CRASH PROGRAM

The PAM-CRASH is a non-linear explicit finite element dynamic analysis program, which, for the puprose of the simulation of continuous fibre reinforced composite crash events, contains the bi-phase and matrix fracturing solid element material model, a bi-phase and matrix fracturing multi-layered composite thin shell model, and a fracturing quasi-isotropic (mono-layer) thin shell model, for the simulation of short random fibre reinforced composite dynamic failure (or crash) behaviour. The bi-phase and fracturing material models are complemented by an internal viscous damping law and by a plastic material behaviour component, which can serve to dissipate energy from high frequency oscillations [61].

The crushable foam, solid brick material model and the special honeycomb sandwich core brick model can serve to simulate foam and honeycomb (cellular) sandwich cores in sandwich FE models, with multi-layered or mono-layered straight fibre, tissue fabric, or random fibre-reinforced thin shell or membrane facings.

The PAM-CRASH tool, with its composite solid, thin shell and membrane material models, is particularly well-suited for dynamic destructive tensile, compression, and bending single wall and sandwich coupon test simulations. Since equivalent "macro-wall" finite elements, calibrated on component crushing tests and/or on detailed PAM-FISS multi-layered wall finite element analysis results, which absorb equivalent amounts of crushing energy, can be constructed, the PAM-CRASH code may be suitable for the simulation of crushed components, sub-assemblies and full composite structures; see also Reference [61].

CHAPTER 4

ENERGY ABSORPTION CAPABILITY OF THIN-WALLED COMPOSITE STRUCTURAL COMPONENTS

4.1 DEFINITION

Most of the studies to examine the energy aborbing capabilities of composite materials have been directed towards the axial crush analysis of composite thin-walled structural components, because the axial crush mode represents more or less the most efficient design. However, impacted structures quite often fail in a mode associated mainly with bending. Therefore, an understanding of the bending crush behaviour of thin-walled composite shells is also necessary and important.

The static axial collapse tests can be carried out between the parallel steel platens of a hydraulic press at very low crosshead speed, whilst the corresponding dynamic ones can be performed by a direct impact on a drop-hammer or an impactor. In an axial loading test, the energy absorbed by the collapsed specimen during the crushing process is calculated by measuring the area under the corresponding load, P/shell shortening (displacement), s curve, see Figure 5.3(b) in Chapter 5. Initially the shell behaves elastically and the load rises at a steady rate to a peak value, P_{max} and then drops abruptly; the magnitude of the peak load is greatly affected by the shell geometry and the material characteristics. As deformation progresses, the shape of the load/displacement curve depends on the mode of collapse and the loading conditions. For thin-walled composite shells subjected to axial collapse, the fracture behaviour of the shell appears to affect the loading stability, as well as the magnitude of the crush load and the energy absorption during the crushing process; shells which collapse in a stable, progressive and controlled manner can dissipate a large amount of energy. The post-crushing region of the load/displacement curve is characterised by high serrations due to the crush energy of the composite material, being absorbed by a sequence of different microcracking processes characterising each case of collapse. The main feature of a curve of a statically loaded shell in the post-crushing region is the characteristic oscillation about a mean post-crushing load, $\overline{P}$ accompanied by shallow serrations, contrariwise to the corresponding dynamic one, which are characterised by successive severe fluctuations with troughs and peaks probably due to the different microfracturing mechanism.

In general, the energy absorption capability of an axially loaded shell of a given material is quantified by the specific energy absorption, W_S. This is defined as the ratio of the energy absorbed, W for the collapsed specimen, see Figure 5.3(b) in Chapter 5, per unit mass crushed, m_C, calculated as the crushed volume, V_C times the material density, ρ.

The bending tests of thin-walled shells can be performed using three or four point loading. A different method proposed by Mamalis et al. [62] can be also used, see the detailed description below. The energy absorbing capability of the shell is calculated, in every case, by measuring the area under the corresponding bending moment, M/angle of rotation, θ curve, see Figure 5.19 in Chapter 5. During the bending process, the shell initially deforms elastically and the M/θ curve is characterised by a sharp, steady-state increase of the bending moment until a maximum value, M_{max} is attained and fracture occurs. The post-crushing regime follows and it is characterised by deep collapse, where the initially developed transverse crack spreads gradually or rapidly over the whole cross section of the shell.

4.2 FACTORS AFFECTING THE ENERGY ABSORPTION CAPABILITY

The findings of the extensive research work, which has been carried out pertaining to the axial collapse and bending of thin-walled structural components, have demonstrated that there are several variables which may control the energy absorption capability of composite materials, the principal ones of which are listed here.

4.2.1 Materials

For the analysis and finally the design of a structure subjected to loading it is necessary to primarily perform the experimental characterisation of the material used. Experimental characterisation refers to the determination of the material properties through tests conducted on suitably designed specimens. Understanding the material response over the entire range of loads is necessary, if advanced design procedures are employed for efficient material utilisation. In the case of composite materials, it may be desirable to begin the design with constituent material properties and arrive at the composite macromechanical properties through micromechanics analyses. High strain-rate or impact loads may be expected in many of the engineering applications of composite materials. The suitability of a composite for such applications is, therefore, determined not only by the usual design parameters, but by its impact or energy-absorbing properties.

The composite material components (fibres and matrix), as well as the laminate design (fibre orientation), greatly affect the crashworthy capacity of structures made of composite material. Fibre content, diameter and length, matrix mechanical properties, as well as the fibre distribution in the laminate, have a significant influence on the energy absorption capability of thin-walled composite shells subjected to axial loading. Moreover, the temperature has an important effect on material properties, with an obvious effect on the material crashworthy response. The most important remarks of the research work that has been done in this topic are reported below.

FIBRE AND MATRIX MATERIALS

The majority of previous studies have concentrated on composites involving fibres of carbon, glass or aramid in a thermosetting resin, such as epoxy. Thornton [63], Farley [64], Schmueser and Wickliffe [65] and Farley and Jones [66,67] all report that, in tests conducted on comparable specimens carbon-epoxy tubes generally absorbed more energy than glass-epoxy or aramid-epoxy specimens. Hybrid composite specimens have been investigated in an attempt to combine the best energy absorption characteristics of different fibres into a single composite material. For example, it might be desirable to combine the post-crushing integrity of aramid with the high specific energy absorption of carbon. However, Farley [64] reports that the energy absorption capabilities of a selection of hybrid specimens were not significantly better than those of single-fibre type with the same ply orientation. Furthermore, Thornton and Edwards [68] report that the presence of aramid in glass-aramid and carbon-aramid hybrids led to unstable collapse by folding, which would not have otherwise occurred had the samples been composed of glass or carbon fibres alone. Farley [69] reports that the energy absorption capability of circular carbon-epoxy tubes is greatly dependent on the strain at failure of both the fibre and the matrix. Of the material systems tested, that with the higher strain at failure exhibited the superior energy absorption capability. Furthermore, it is suggested that to obtain maximum energy absorption from a particular fibre, the matrix material in the composite must have a higher strain at failure than the fibre.

New fibre and matrix materials continued to be employed in attempts to obtain ever higher values for specific energy absorption. Reference [70] reports on a Dyneema PE fibre/carbon fibre hybrid composite tube, which showed a high specific energy absorption. Even more impressively, Hamada et al. [71] describe the use of a thermoplastic polyetheretherketone (PEEK) matrix with a carbon fibre, which gave an exceptionally high specific energy absorption value of 180 kJ/kg (at least double the value of carbon-epoxy). This is attributed to the PEEK matrix offering a high resistance to crack growth between the fibres, preventing failure by this mode until the onset of stable progressive crushing. However, according to Gosnell [72] the cost of thermoplastic resins and prepregs is relatively high in comparison to thermosets and Nilson [73] notes that, in processing PEEK one can encounter high pressures and temperatures and difficulty with fibre wet-out. Nilson also reports, there are still many aspects of PEEK to be fully investigated including notch sensitivity, the strength of bonded and bolted joints, the effect of temperature on stiffness and aspects of fatigue.

LAMINATE DESIGN

Thornton and Edwards [68] report that cohere tubes with a $(45/45)_n$ lay-up developed consistently lower values of specific energy than tubes with $(0/90)_n$ lay-ups in the stable collapse region. Farley [64] reports that, energy absorption capability tends to vary with ply orientation. Considerable variations in energy absorption capability were observed in quasi-static tests on $[0/\pm\theta]$ carbon-epoxy circular tubes for $0° < \theta < 45°$ with energy absorption decreasing as θ increased. Smaller variations in energy absorption capability were also observed in $[0/\pm\theta]$ aramid-epoxy and $[0/\pm\theta]$ glass-epoxy circular tubes for $45° < \theta < 90°$, with specific energy absorption generally increasing with increasing θ. These findings are broadly supported by Kindervater [74]. Schmueser and Wickliffe [65] also report variations in specific energy absorption with ply orientation, although the nature of the variation differs from that described by Farley. They report that the specific energy absorption of carbon-epoxy, glass-epoxy and aramid-epoxy $[0_2/\pm\theta]$ specimens generally increases with increasing θ. However, Schmueser and Wickliffe's tests appear to differ from those of Farley's in that they were dynamic tests conducted in a drop hammer and, unlike Farley, they did not appear to have used a collapsed trigger mechanism (see below) to initiate stable high energy collapse.

Mamalis et al. [75,76], in an extensive experimental work on the axial collapse of thin-walled circular and square tubes made of two different composite materials, report that the specimens made of a commercial glass fibre and vinylester composite material, consisting of nine plies that were laid-up in the sequence $[(90/0/2R_c)/(2R_c/0/90)/R_{c.75}]$, show better energy absorbing characteristics than those made of a fibreglass composite material in which the glass fibres were in the form of chopped strand mat with random fibre orientation in the plane of the mat. Furthermore, Mamalis et al. [77] report that the existence of extra $(-+-45°)$ plies in the central regions of the hourglass cross-section shells with the sequence $[(90/0/2R_c)/(2R_c/0/90)/R_{c.75}]$ has a great effect on their energy absorbing behaviour, causing a decrease on the specific energy absorbed.

TEMPERATURE

Elevated temperatures affect the crush characteristics of composites, primarily through changes in resin properties. Thornton [63] reports that the specific energy absorption of carbon and glass composite tubes generally decreases quite substantially with increasing temperature above about 0°C, see Figure 4.1.

4.2.2 Structural Geometry

Extensive analytical work was performed by a great number of researchers, pertaining to the crashworthy behaviour of fibreglass composite thin-walled structural components of various geometries. The effect of specimen geometry on the energy absorption capability was investigated varying the various geometric parameters of the shells, such as wall thickness, t, axial length, L, mean diameter, $\overline{D}$, or circumfer-

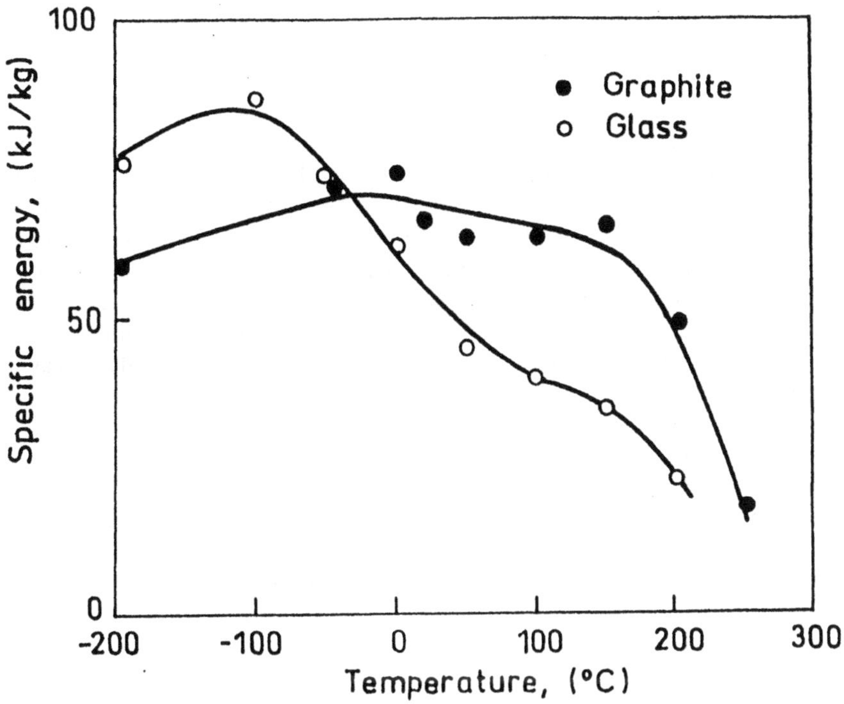

Figure 4.1. Variation of specific energy with temperature for graphite and glass composite tubes.

ence, C and semi-apical angle, θ in the case of frusta. Moreover, the trigger mechanism used was also investigated to indicate the effect on the shell energy absorbing capacity.

CIRCULAR TUBES

Farley [78] reports that energy absorption capability is a non-linear function of diameter to thickness, D/t ratio for $[\pm45]_n$ carbon-epoxy and aramid-epoxy circular tubes. Energy absorption was found to fall as D/t increases. This finding is also supported by Farley and Jones [79]. Furthermore, Farley [78] reports that the carbon-epoxy specimens exhibited different non-linear dependencies on D/t for specimens of different internal diameters. The aramid-epoxy specimens of all diameters exhibited no such diametrical dependence. This implies that aramid-epoxy tube specimens can be geometrically scaled for energy absorption, whilst carbon-epoxy tubes cannot. Thornton and Edwards [68] report that critical values of relative density (a quadratic function of thickness to diameter ratio) can be identified above which (in the case of carbon and glass FRP tubes) or below which (for aramid FRP tubes) stable collapse occurs with high energy absorption. Mamalis et al. [63], in an investigation concerning the static axial loading of glass polyester circular tubes, used the tube wall slenderness ratio, $t/\overline{D}$ (t is the wall thickness and $\overline{D}$ the mean diameter of the tube), as a parameter to examine the variation of the total energy absorbed, W, indicating that the energy absorbed increases with increasing $t/\overline{D}$.

To approach the square-wave type of load/displacement response, it is necessary to ensure that neither brittle fracture nor any form of buckling instability can occur. It has been established that this cannot be achieved with tubes having square ends, and a trigger mechanism must, therefore, be used to promote some form of progressive deformation. Farley [64] reports that modifying one end of the tube, e.g. by introducing a chamfer, can greatly reduce the peak load experienced by the specimen without affecting the sustained crushing load, see Figure 4.2. Tulip type triggers have also been shown to perform well. Triggering is less important in aramid composite tubes, because they tend to fail by local buckling in a similar manner to metals such as aluminium.

SQUARE/RECTANGULAR TUBES

Thornton and Edwards [68] report that the square and rectangular cross-section tubes are generally less effective at absorption energy than circular ones. These findings are also supported by Mamalis et al. [76, 80] and Kindervater [74], who report

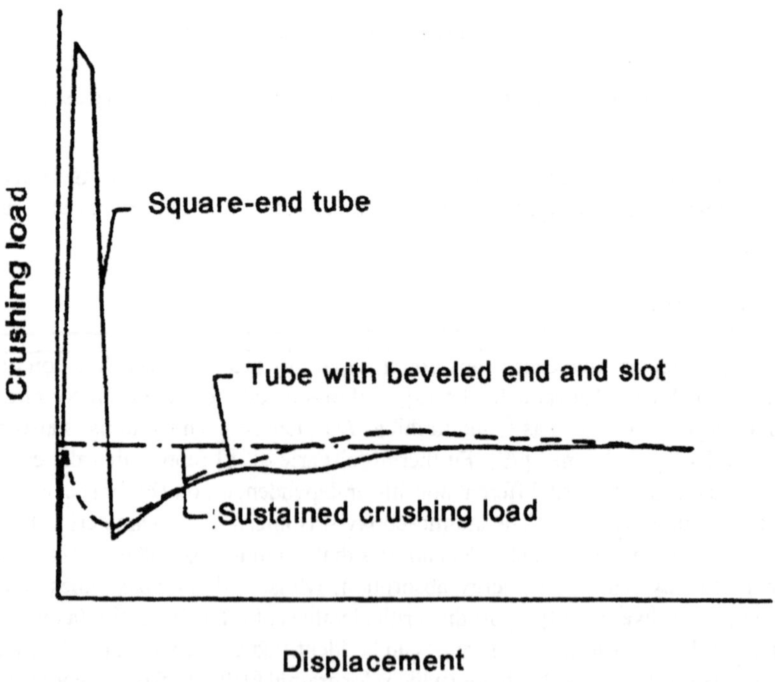

Figure 4.2. The effect of collapse trigger mechanism on the load/displacement curve of a composite tube subjected to axial collapse.

that square and rectangular cross-sectioned tubes have respectively 0.8 and 0.5 times the specific energy absorption of comparable circular specimens. The reason for the lower specific energy absorption of square and rectangular sections, is generally attributed to the fact that the corners act as stress concentrations leading to the formation of splitting cracks there. This tends to result in unstable collapse with low energy absorption. Furthermore. Mamalis et al. [76] report that the axial length of square tubes subjected to axial loading and collapsing in a stable manner does not affect their crashworthy capability.

Czaplicki et al. [81] report that the energy absorbed by tulip triggered specimens was significantly higher than for bevel triggered ones of the same geometry and material. In addition, the crushing was more controlled and predictable with the tulip trigger.

CONICAL SHELLS

Mamalis et al. [75,82] report a general trend that specific energy absorption decreases as the semi-apical angle of the frusta increases. This supports the findings of an earlier paper concerning circular frusta, see Reference [83], in which a transition point between stable and unstable collapse due to the effect of semi-apical angle is identified as laying between 15 and 20 degrees. Furthermore, it was observed that, contrary to what is commonly experienced with the collapse of cylindrical tubes, conical specimens do not require a collapse trigger mechanism to avoid initial catastrophic failure. Mamalis et al. [82] also report that the crashworthy characteristics of the square frusta axially loaded underestimate the corresponding ones concerning circular frusta due to the same reason mentioned above, explaining the less crashworthy capability of square tubes than circular ones.

OTHER GEOMETRIES

A non-conventional hourglass cross-section automotive frame rail, see Figure 9.1(a) in Chapter 9, made of a glass fibre/vinylester composite material has been designed for use in the apron construction of the car body. The crashworthy behaviour of this structural component in axial collapse has been studied by Mamalis et al. [77]. In this work is reported that, the specific energy of the progressive collapsed specimens seems to be almost constant as the geometry factor thickness/axial length, t/L of the shell increases, supporting the remark that for a constant thickness of the shell, its axial length has no significant effect on its energy absorbing capability. Furthermore, it is also reported that hourglass sections developed higher values of specific energy than square tubes made of the same material and loaded under the same conditions.

Farley and Jones [79] describe the effect of reducing the included angle of "near-elliptical" carbon-epoxy tubes, see Figure 4.3, on their energy absorption capability. As it can be seen in that figure, the tubes actually consist of two identical parts of a circular tube where the centres of curvature of the two parts are not coincident. The greater the distance between the two centres, the smaller the included an-

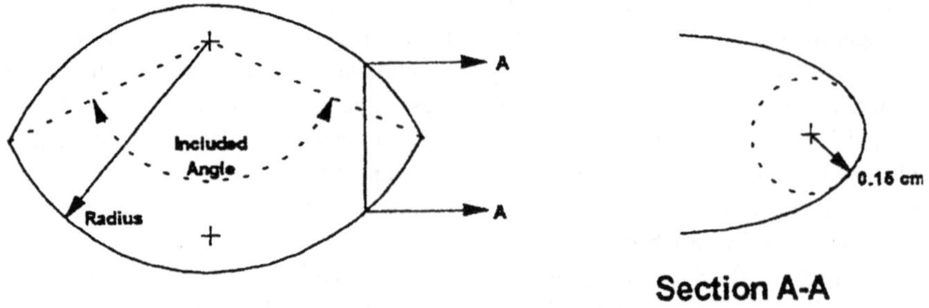

Figure 4.3. Cross-section of a "near-elliptical" shell.

gle. To facilitate fabrication and minimise fibre damage at the ends of the major axis, a 15 mm radius was incorporated. For both material systems, the energy absorption capability increased as the included angle decreased, i.e. the energy absorption capability increased as the tubes became "more" elliptical. As the included angle was reduced from 180° (circular) to 90°, the energy absorption capability increased by between 10% and 30%. This is attributed to an increased percentage of material at the ends of the major axis, where there is a locally reduced diameter to thickness ratio and, hence, increased energy absorption capability. The crushing mode near the ends of the major axis was predominantly high energy absorption brittle fracturing, whilst that away from these regions was mainly lower energy absorption lamina bending.

4.2.3 Loading

AXIAL LOADING

The influence of strain-rate on the energy absorbing capacity of composite shells subjected to axial loading was investigated by many researchers. Significant experimental work has been carried out on static and dynamic crushing, but conflicting results were produced. Energy absorption dependence on crush-speed is related to the mechanism which controls the crushing process; if the mechanism is a function of strain-rate then the energy absorption of the tube will be a function of crush-speed. In general, strain-rate and, therefore, crushing speed may influence the mechanical properties of the fibre and the matrix of the composite material. Moreover, the friction coefficients between the various slided surfaces during the crushing process may be influenced by changes in crush-speed.

Farley [84] reports that matrix stiffness and failure strain can be a function of strain-rate and, therefore, the energy absorption, associated with interlaminar crack growth, may be considered as a function of crush-speed, whilst, the mechanical properties of brittle fibres are generally insensitive to strain-rate and, therefore, the fracturing of the lamina bundles generally is not a function of crush-speed. In this paper, Farley is also dealing with the axial loading of circular tubes, reporting that the energy absorption capability of the tested tubes is influenced by crush-speed. Specimens were crushed at constant crush-speeds of between 0.01 m/s and 12 m/s; $[\pm\theta]_3$ carbon-epoxy specimens were found to be a weak function of crush-speed, with an increase in energy absorption of as much as 35% (depending on ply orientation) over the speed range tested. Furthermore, all aramid-epoxy specimens tested were found to be a function of crush-speed, with energy absorption increasing between 20% and 45%. However, the $[0/\pm\theta]_2$ carbon-epoxy tubes were not found to be a function of crush-speed. From the remarks pointed out above, concerning the governing mechanism which controls the crushing process, it may be assumed that for the $[0/\pm\theta]_2$ carbon-epoxy specimens, the fibre (whose machanical properties are not a function of strain-rate) controls the crushing process, whilst for the $[\pm\theta]_3$ carbon epoxy-tubes, the matrix (whose mechanical properties are a function of strain-rate) is the primary influence.

Mamalis et al. [75, 80, 82] report that friction mechanisms, which are developed between the composite material and the crushing surface and between the various "new" surfaces which have formed after interlaminar crack growth (adjacent lamina bundles), are also affected by the strain-rate. Furthermore, they report that, the crashworthiness results during dynamic axial collapse are lower than the static ones for thin-walled fibreglass/polyester circular and square tubes and frusta.

Kirch and Jannie [85], based on a test series of glass polyester tubes of various section geometries impacted under a drop-hammer at a speed of 13 m/s, show that the strain-rate effect on the specific energy is the essential one. Schmueser and Wickliffe [65] in a similar work on $[0_2/\pm\theta]_s$ circular tubes made of epoxy resin and graphite, glass and Kevlar fibre, respectively, indicate that the dynamic specific energy is lower than the static one for all three types of composite material examined. On the contrary, Mamalis et al. [76, 77] report that dynamic collapse overestimates static collapse by about 20%, as far as the crashworthy characteristics of fibreglass/vinylester composite shells of various geometries is concerned, probably due to higher values of the dynamic friction coefficients. Similar observations are reported by Berry and Hull [86], with the specific energy increasing with increasing loading rate up to 8.5 m/s for the $[0/90]_n$ graphite/epoxy and $[0/90]_n$ glass/epoxy composites. Different remarks are reported by Thornton [87], where the specific energy of the $[0_2/\pm\theta]_n$ graphite/epoxy composite is not a function of the crush-speed, whilst the $[\pm\theta]_3$ graphite/epoxy composites exhibit an increase up to 30% at 12 m/s; energy absorption increases between 20% and 45% for both $[0_2/\pm\theta]_2$ and $[\pm\theta]_3$ kevlar/epoxy composites. Kindervater [74] suggests that the effect of crush-speed on energy absorption will vary, depending on the particular material system used. Carbon-epoxy tubes showed as much as a 20% degradation in energy absorption capability under impact loading of up to 9 m/s, whilst high performance polyethylene fibre Dyneema SK60 in an epoxy matrix and carbon fibres embedded in a thermoplastic polyamid matrix, showed increases of approaching 50%.

There is the possibility that in real structures the mode of failure may be very different in static and dynamic instances. Savage [12] reports that in quasi-static tests conducted on the composite nose cone of a Formula One racing car, the failure machanism was one of global buckling of the composite skins rather than the progressive crushing and high energy absorption associated with dynamic impact of the cone. Care should, therefore, be exercised when using static test data to predict dynamic behaviour.

BENDING

A literature survey shows that very little work has been reported on the collapse of composite thin-walled tubular components made of composite material due to bending. Mamalis et al. [33, 62, 88] report on the bending of thin-walled fibre-reinforced glass vinylester and polyester composite tubes of various geometries under certain end-clamping conditions, simulating the oblique collision of stuctural elements of impacted vehicles. The test series performed in a self-made experimental set-up. The specimen was suitably clamped at one end and supported at a point close to its opposite end, whilst the torque required for tube bending was supplied by a speed reducer driven by an electric motor.

Reference [62], examining the bending of circular composite tubes, reports that the region of the test specimen under tension exhibited different fracture characteristics to that under compression. Furthermore, the tube clamping mechanism was found to greatly affect the amount of energy absorbed. The insertion of a plug at the clamped end was found to lower energy dissipation due to shortening of the post-crushing regime, whilst clamping devices with rounded edges tended to delay crack development and propagation, resulting in generally higher bending moments and energy absorption. Both the insertion of plugs and the use of rounded clamping devices was found to increase the peak bending moment exhibited by the specimens.

Reference [33] extended the above study to the bending of square and rectangular cross–sectioned tubes. The peak bending moment and the energy absorption of both square and rectangular composite tubes were found to increase with increasing wall thickness, with the rectangular specimens exhibiting the better energy absorption when bent over their strong axis. In general, rectangular tubes were found to show better crashworthy characteristics for large deformations under bending than circular tubes of comparable dimensions. This was attributed to the deformations undergone at the corners of the tube and the increased tube strength under bending.

Reference [88] moved away from the bending of tubes with simple geometrical cross sections to a much more complex hourglass shape, similar to that described previously and shown in Figure 9.1(a) in Chapter 9. As mentioned above, this structure had been designed for use in the apron location of a car body to provide a high degree of crashworthiness. The effect on energy absorption of plug inserts at both clamped and free ends of the tube was examined and, in contrast to the case for the circular tubes described above, the plugs were found to have a positive effect in lengthening the post-buckling region but not to affect the peak bending moment. Just as for the rectangular tubes described in Reference [33], bending over the strong axis of the beam resulted in better energy absorption capability.

Mahmood et al. [89] report on the bending of composite shells with hourglass and rectangular cross-sections with purpose of the characterisation of their crush strengths.

Two types of test arrangement were used, i.e. three and four point loading. The observations showed that the mode of collapse of the beams suffers more damage in the compression zone than that in the tension zone. At the maximum strength, the applied compressive stress is mainly resisted by the edge of the sub-element or the corners of the component. However, if the corners suffer material separation this leads to a sudden loss of bending resistance in deep collapse and poor energy management.

COMBINED LOADING

In a head-on collision the various structural components do not collapse in a simple, ideal form but in a non-axial manner. Non-axial crush means that components are subject to combined axial and bending loads. Czaplicki et al. [90] report an investigation of two types of non-axial crushing, i.e. off-axis crushing and angled crushing. Angle loading occurs when a vehicle, moving forward along its longitudinal axis, impacts an object tilted away from being perpendicular to the vehicle's longitudinal axis, whilst off-axis loading occurs when a spinning vehicle impacts an object from a direction not along its longitudinal axis. The disparity in behaviour between angled and off-axis loading results from differences in the friction between the tube and the crushing plate. Off-axis loading involves a dynamic type of friction, whilst angled loading a static one. Furthermore, it is reported that significant differences appeared between the E-glass/polyester pultruded tubes crushed in these different configurations. The energy absorption for the two types of crushing is found also to vary significantly with the angle of inclination, with the difference being largest at high angles of inclination, and that, in general, the energy absorbed is less in off-axis than in angled loading.

4.3 FAILURE MECHANISMS/MECHANICAL RESPONSE

4.3.1 Macroscopic Collapse Modes

AXIAL LOADING

Thin-walled structural components of various simple geometries made of composite materials and subjected to axial loading were found to collapse in modes considerably different than those observed in metallic and thermoplastic structures. The brittle nature of both fibres and resin ensures that composite materials do not undergo the plastic deformation, characteristic for ductile metals and PVC; see Mamalis et al. [8] for more details about the crashworthy behaviour of metallic and PVC thin-walled structural components. Contrariwise, the dominant mechanism in the present case is that of fracture and fragmentation. The failure modes observed throughout these series of tests are, in general, greatly affected by the shell geometry, the arrangement of fibres, the properties of the matrix and fibres of the composite material and the stacking sequences.

The macroscopic collapse modes of thin-walled composite shells subjected to low speed axial loading may be classified as: (a) stable progressive collapse modes, associated with controlled crushing process, and (b) unstable ones, associated with extensive brittle fracture. Note that the whole crushing process and, therefore, the macroscopic collapse modes of the axially loaded composite shells greatly affect the energy absorbing capability of the structural components. Based on the experimental observations from a very extensive experimental treatment of axisymmetric tubes of various geometries made of fibre-reinforced polymer matrix composite materials by Mamalis et al. [76, 77, 83, 91, 92], the following main modes of failure may be identified and classified:

- Progressive crushing with microfragmentation of the composite material associated with large amounts of crush energy, designated as Mode I. Three different modes of failure were observed: Mode Ia of failure, similar to a "mushrooming" failure; Mode Ib of collapse, which is characterised by the inversion of the shell wall inwards; Mode Ic, which is characterised by an outwards inversion of the shell wall.
- Brittle fracture resulting in catastrophic failure of the shell with little energy absorption, designated as Mode II and Mode III, depending on the crack form.
- Progressive folding and hinging, similar to the crushing behaviour of thin-walled metal and plastic tubes, showing a medium energy absorbing capacity, designated as Mode IV.

Furthermore, the experimentally obtained deformation modes of all specimens tested are classified in respect to the geometry factors: wall thickness/mean circumference, *t/C* and axial length/mean circumference, *L/C*. Distinct regions, characteristic for the various deformation modes developed, and the transition boundaries from stable to unstable modes of collapse are indicated, providing, therefore, useful information about the collapse of the various geometries and their behaviour as energy absorbers.

Axial loading of various shell geometries at elevated strain-rates has been also undertaken by Mamalis et al., see References [76, 82, 92]. The modes of collapse observed can be classified as stable and unstable collapse modes; stable collapse modes show similar features to those obtained during the static loading of the same geometries.

BENDING

The collapse modes at macroscopic scale of various cross-section composite thin-walled shells, such as circular, square and rectangular tubes and hourglass cross-sectioned shells, subjected to bending under certain end-clamping conditions are reported by Mamalis et al. [33, 62, 88].

In Reference [62] the collapse behaviour of bent circular tubes is examined; two distinct regions with different macroscopic characteristics were observed, an upper zone subjected to compressive loading and a lower one under tensile straining. Note, also, that a narrow transition zone between the compression-tension regions with combined features was also observed. In general, collapse initiates in the compressive zone, close to the clamping device.

Bending of square and rectangular cross-section thin-walled tubes is reported in Reference [33]. Three distinct regions with different macro- and microscopic characteristics are observed, the top wall of the tube subjected to compression, the bottom wall subjected to tensile straining, whilst the side walls of the tube show combined compression/tension features. Note that the four corners of the rectangular tubes greatly influence the above mentioned mechanisms.

Similar remarks concerning the macroscopic collapse modes are also reported in Reference [88], where the bending of hourglass cross-sesctioned shells is examined. Moreover, it is reported that for bending about the major axis failure occurs at the top compression zone, with the bottom tension zone experiencing no failure even after the contact region at the loading zone is collapsed entirely. Cracking along the middle surface is also observed. For bending about the minor axis, failure again occurs at the top compression zone and the top corners are sheared, whilst the tensile zone remains intact. No cracking is observed along the middle surface. Mahmood et al. [89], also examined the bending characteristics of various thin-walled shells, i.e. square and rectangular cross-sectioned tubes and rail beams (hourglass cross-sectioned shells). Although they used three and four point loading, they report similar remarks with Mamalis et al. [33,88] as far as the collapse modes is concerned; the only difference is that the failure occurs at the top compression zone under loading nose(s).

4.3.2 Microfracturing

In a very broad sense, failure of a structural element can be stated to have taken place when it ceases to perform satisfactorily. Therefore, the definition of failure may change from one application to another. In some applications a very small deformation may be considered failure, whereas in others only total fracture or separation constitutes failure. In the case of composite materials, internal material failure generally initiates much before any change in its macroscopic appearance or behaviour is observed. The internal material failure may be observed in many forms, separately or jointly, such as breaking of the fibres, microcracking of the matrix, separation of fibres from the matrix in the form of debonding or/and pull-out and separation of laminae from each other in a laminated composite (called delamination). The effect of internal damage on macroscopic material response is observed only when the frequency of internal damage is sufficiently high.

AXIAL LOADING

Farley and Jones [66] report that the crushing response of composite tubes can be classified into three basic modes: transverse shearing, lamina bending and local buckling. The transverse shearing and lamina bending crushing modes are exhibited exclusively by brittle fibre-reinforced composites, whilst both ductile (such as Kevlar) and, in some cases, brittle fibre-reinforced composite materials can exhibit the local buckling crushing mode, similar to that exhibited by ductile metals. Moreover, they report that, when a load is applied to the edge of the crushing initiator, local failure of material occurs and small inter/intralaminar cracks are formed, where their length determines, whether the resulting crushing mode is transverse shearing, lamina bending or a combination of these modes (brittle fracture). In the case of the transverse shearing crushing mode, the lengths of the interlaminar and longitudinal cracks are typically less than the thickness of the laminate, whilst the lamina bending crushing mode is characterised by very long interlaminar, intralaminar and parallel-to-fibre cracks; their lengths are greater than ten laminate thicknesses. No fracturing of lamina bundles occurs. In the case of the brittle fracturing crushing mode, the length of the interlaminar cracks varies between one and ten laminate thicknesses.

The local buckling crushing mode consists of the formation of local buckles by means of plastic deformation of the material. The post-crushing integrity of ductile fibre-reinforced composites is a result of fibre and matrix plasticity, i.e. significant deformation without fracture and fibre splitting. Brittle fibre-reinforced composites exhibit the local buckling crushing mode when, (i) the interlaminar stresses are small relative to the strength of the matrix, (ii) the matrix has a higher failure strain than the fibre and (iii) the matrix exhibits plastic deformation under high stress. The mechanisms that control these different crushing modes are a function of the mechanical properties of the constituent materials and the structure of the specimen. In the case of transverse shearing, interlaminar crack growth and lamina bundle fracture are the crushing mechanisms, whilst inter/intralaminar crack growth and friction are the mechanisms in the lamina bending crushing mode. The mechanisms that control the crushing process in the local buckling crushing process are plastic yielding of the fibre and/or the matrix. Crushing response of composite tubes can be a function of crushing speed, provided that the mechanical properties of the mechanisms that control the crushing process are strain-rate sensitive.

Fairfull and Hull [93] report a conceptual model of the crush-zone configuration, based on an analysis concerning the microfracture mechanism of thin-walled circular tubes subjected to static axial loading and following the progressive collapse mode (Mode Ia), see Figure 4.4. Similar remarks are reported by Mamalis et al. [75, 91], see also Figure 5.6 in Chapter 5. Furthermore, Mamalis et al. [76, 77] report an extensive description of the microfracture mechanism of square tubes and hourglass sections axially loaded, which is similar to that obtained for circular tubes. The main features of this microfracture mechanism are:

- An annular wedge of highly fragmented material, forced down axially through the shell wall
- an intrawall microcrack, which develops ahead of the crush-zone at the apex (tip) of the annular wedge and propagates at a rate approximating the compression rate
- two continuous fronds (internal and/or external) as a result of the plies delamination in the crush zone, mainly caused by the central bundle wedge, which spreads radially inwards and outwards from the wall of the frustum

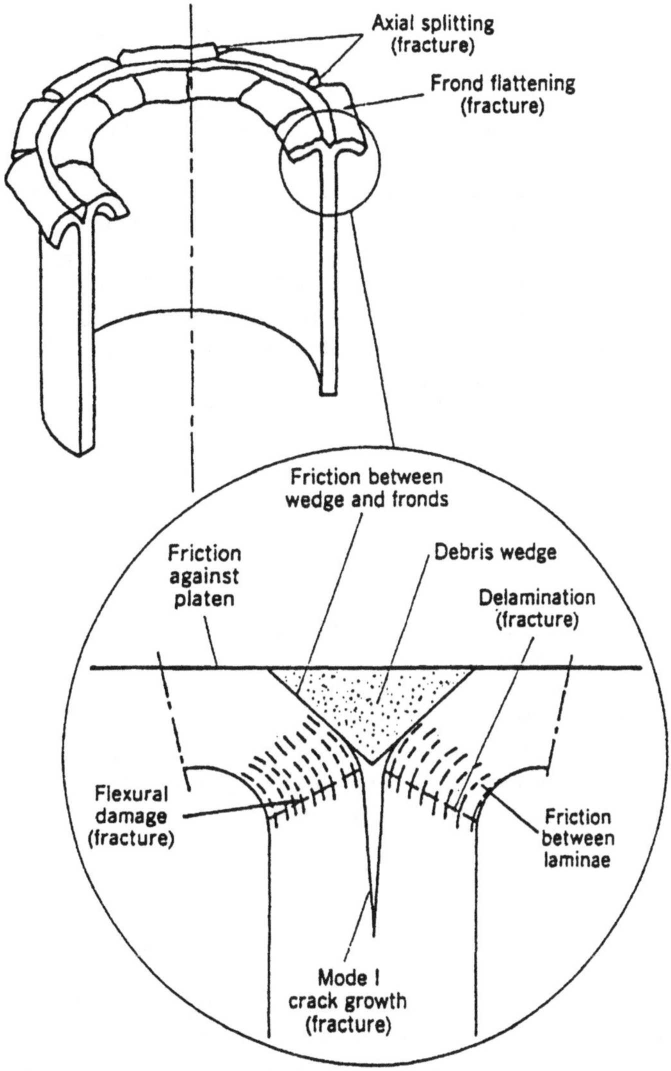

Figure 4.4. A schematic diagram of the failure mechanism of thin-walled composite tubes subjected to axial loading.

- a severely strained zone (compressive-tensile zone), which extends between the central crack and the shell wall edges showing a combined tensile-compressive type of deformation

Regarding the microfracture mechanism of the circular and square frusta subjected to axial loading, as far as the Mode Ia of collapse is concerned, Mamalis et al. [92, 94]

report that the experimental observations made are similar to those obtained during the axial collapse of tubes, but some characteristic differences are obtained. Since the circumference of the shell increases as the crushing of the frusta progresses, it is evident that the size of the wedge increases during crushing. With increasing semi-apical angle of the frustum, the position of the intrawall crack moves towards the outside edge of the shell wall, increasing in this manner the thickness of the inner frond and simultaneously resulting in a positioning of the annular wedge mainly above it, see Figure 7.10 in Chapter 7. On the contrary, the crack length decreases with increasing semi-apical angle. Furthermore, Mamalis et al. [82, 92] report that the microfracture mechanism for the progressive collapse of circular and square frusta subjected to dynamic loading is, in general, similar to that obtained during the axial static collapse; the only differences encountered are related to the shape and the position of the wedge and the microcracking development.

Based on the above mentioned micromechanism, as well as on secondary failure mechanisms contributing to the overall energy absorption during collapse as suggested by Fairfull and Hull [93] and also supported by Mamalis et al. [75], from the energy aborption point of view, the following principal sources of energy dissipation at microscopic scale may be listed:

- Intrawall crack propagation
- Fronds bending due to delamination between plies
- Axial splitting between fronds
- Flexural damage of individual plies due to small radius of curvature at the delamination limits
- Frictional resistance to axial sliding between adjacent laminates
- Frictional resistance to the penetration of the debris wedge
- Frictional resistance to fronds sliding across the platen

BENDING

Mamalis et al. [62] report a detailed description of the micromechanism failure of circular tubes subjected to bending. Two distinct regions, i.e. a top compressive zone and a bottom tensile zone, with different microscopic characteristics are developed during the bending process. The main features of the regions under compressive loading are characteristic cracking of fibres and resin as well as bending and/or buckling of longitudinal fibres without fracture. On the contrary, tube regions loaded in tension exhibit complex heterogeneous damage and failure modes including extensive fibre/matrix interface debonding and delaminations.

Mamalis et al. [33, 88] describe the microscopic failure of bent non-circular components, such as square and rectangular tubes in the first paper and hourglass cross-sectioned shells in the second one. They report that in both cases the cracking characteristics may be classified into three failure modes, i.e. flexural, delamination buckling and shear mode, depending on the loading conditions, the strained region and the cracking development. In the case of the flexural mode, the typical damage is fibre and matrix breakage on either side of the laminate. Delamination buckling, which is the dominant failure mode, is characterised by a delamination crack developed usually in the middle of the laminate, see Figure 6.13(a) in Chapter 6, whilst in failure by shear the typical damage is fibre and matrix breakage through the thickness of the shell.

4.4 PREDICTIVE TECHNIQUES

The destructive nature of the experimental energy absorption tests described above would suggest that the cost of testing may become a significant consideration in an extensive investigation. Indeed, it is almost certain that there will be financial constraints upon large-scale energy absorption/crashworthiness tests, such as those conducted using full-size vehicles. Therefore, any methods, which can successfully and relatively cheaply predict the energy absorption capability of composite materials, are clearly welcomed.

4.4.1 Failure Analysis

Fairfull and Hull [93] report a conceptual model estimating the load fractions undertaken by the different regions of the crush zone of axially loaded tubes using the results of various measurements, whilst Yuan and Viegelahn [41] report on modelling of the crushing behaviour of fibreglass tubes, based on the results of various tests, such as tension, compression, fracture toughness and bending.

Farley and Jones [67] describe a simplified procedure, based on an equation similar to the buckling load equation for a column on an elastic foundation, which can be used to predict whether a change in energy absorption of a composite tube occurs as a result of changes in geometry or material properties.

Mamalis et al. [95] report a theoretical analysis of the stable collapse mechanism of the conical thin-walled shells, crushed under axial static and dynamic compression, for calculating the energy absorbed during collapse. The analysis is based on experimental observations taking also into account all possible energy absorbing mechanisms developed during the process.

Similar theoretical predictions of the energy absorbing capacity of other shell geometries, i.e. circular tubes, square tubes and frusta and hourglass sections, are reported by Mamalis et al. [75–77, 92].

Mamalis et al. [96] describe the deformation mechanism of thin-walled composite tubes of non-circular cross section subjected to a cantilever bending. Furthermore, they report on the theoretical analysis for the prediction of the ultimate bending strength for tubes made of various composite materials and cross sections. The proposed analysis provides also the designer with the abilty to describe the bending moment, M/angle of rotation, θ curve in the elastic regime.

4.4.2 Numerical Simulation

Farley and Jones [97] describe the formulation of a finite element model based on a phenomenological approach to the crushing process in order to predict the energy absorption of aluminium, carbon-epoxy and aramid-epoxy tubes. The strain energy release rate and the maximum failure strain were used as criteria for interlaminar crack growth and fracture of laminar bundles. The finite element representation employed was a one-quarter symmetric four-layer model with one layer for each ply of the composite tube. Mathematical zero-length springs were used to connect adjacent layers and the interlaminar crack growth was represented by removal of these springs.

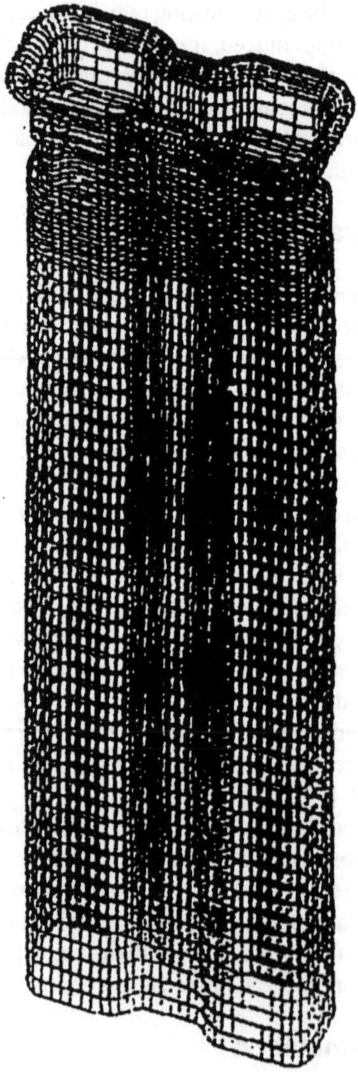

Figure 4.5. FE modeling of the deformation mechanism of an hourglass shaped section subjected to axial loading.

Agreement between predicted and measured energy absorption values was to within 14%, 25% and 28% for the aluminium, aramid and carbon tubes, respectively.

The need for further improvement of the model is acknowledged. Haug et al. [61] describe the material models available on the finite element packages PAM-FISS and PAM-CRASH, which appear most suited to the modelling of composite material failure. The use of these codes in predicting the crushing response (including energy absorption) of axially compressed box columns made of hybrid carbon-aramid sandwich panels is described and the numerical model is compared with experimentally obtained data. Botkin et al. [98] describe the use of another finite element composite material failure model, this time implemented in the LS-DYNA3D software. A square thin-walled tube with sharp corners, a square thin-walled tube with rounded corners and a beam with an hourglass shaped section were all modelled, see Figure 4.5, and the results were compared with experimental data. The models were found to provide a reasonable means of predicting trends in the crush performance of the specimens although they did not match the experimental curves very closely in some instances.

4.5 QUANTITATIVE DATA

The energy absorption capability of a crushed material is commonly quoted in the form of its specific energy absorption value as defined at the beginning. Figure 4.6 and Table 4.1 relate some of the specific energy absorption values, which have been quoted in the literature and it can be seen that, in some cases, an extremely wide range of values for a given material are presented. However, this does not mean to say that any values are particularly right or wrong; it merely re-emphasises the fact that there are many factors, which control the energy absorption capability of composites and that only when comparing like-with-like it is possible to obtain a ranking of material energy absorption capabilities [9].

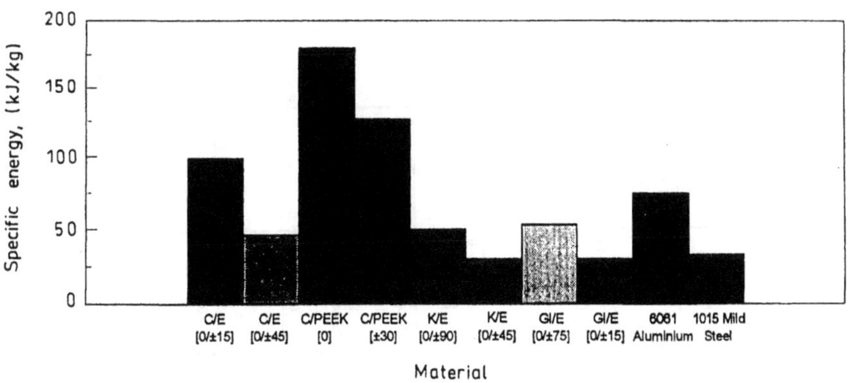

Figure 4.6. Energy absorption capability of various composite materials.

Table 4.1: Comparison of material energy absorption capabilities

Material	Specific Energy Absorption kJ/kg									
	Thornton & Magee 1977 [114]	Thornton 1979 [83]	Thornton & Edwards 1982 [68]	Farley 1983 [64]	Farley 1986 [78]	Farley 1986 [69]	Farley 1991 [84]	Farley & Jones 1992 [79]	Hamada et al. 1992 [71]	Mamalis et al. 1992 [33]
Carbon-Epoxy [0/±15]				99 (t/D = 0.033)		75 (t/D = 0.045)				
Carbon-Epoxy [0/±45]				46 (t/D = 0.032)		55 (t/D = 0.051)	65 (t/D = 0.020)			
Carbon-Epoxy [±45]			55 (t/D ≈ 0.062)		55 (t/D = 0.036)	58 (t/D = 0.048)	50 (t/D = 0.021)	60 (t/D ≈ 0.031)	53 (t/D = 0.05)	
Carbon Fibre - PEEK [0]									180 (t/D = 0.050	
Carbon Fibre - PEEK [±30]									127 (t/D = 0.05)	
Aramid-Epoxy [0/±15]				31 (t/D = 0.036)			9 (t/D = 0.020)			
Aramid-Epoxy [0/±45]				30 (t/D = 0.033)			21 (t/D = 0.022)			
Aramid-Epoxy [±45]		60 (t/D ≈ 0.066)	31 (t/D ≈ 0.042)		33 (t/D = 0.035)		23 (t/D = 0.022)	22 (t/D ≈ 0.030)		
Aramid-Epoxy [0/±90]				50 (t/D = 0.033)						

Table 4.1 (cont.)

Material	Specific Energy Absorption kJ/kg									
	Thornton & Magee 1977 [114]	Thornton 1979 [63]	Thornton & Edwards 1982 [88]	Farley 1983 [64]	Farley 1986 [78]	Farley 1986 [80]	Farley 1991 [84]	Farley & Jones 1992 [79]	Hamada et al. 1992 [71]	Mamalis et al. 1992 [33]
Glass-Epoxy [0/±15]				30 (t/D = 0.060)						
Glass-Epoxy [0/±45]			30 (t/D = 0.028)	31 (t/D = 0.059)						
Glass-Epoxy [0/±75]				53 (t/D = 0.070)						
Glass-Vinylester [90/0/2Rc/2Rc/0/0/90/Rc.75]										54 (t/D = 0.05)
Glass-Polyester $[Rc]_n$										54 (t/D = 0.162)
6061 Aluminium	75 (t/D = 0.060)	72 (t/D = 0.060)		90 (t/D = 0.057)						
1015 Mild Steel	33 (t/D = 0.060)	33 (t/D = 0.060)								

CHAPTER 5

CIRCULAR TUBES

5.1 NOTATION

b_{cr} = angle of axial split
$\overline{D}$ = mean diameter of tube
D_i = inside diameter of tube
D_o = outside diameter of tube
F_x = horizontal force component
G = fracture toughness
k = constant
L = axial length of tube
L_c = length of central crack
l_s = side length of wedge
M = bending moment
M_{max} = peak bending moment
n = number of splits
P = current crushing load
P_1, P_2, P_w = normal force per unit length
$\overline{P}$ = mean crushing load
P_{max} = peak load
R_{ad} = fracture energy per unit area of layers
s = displacement, shell shortening, crush length
T = shear force per unit length
t = wall thickness of circular tube
v = crush-speed
W = energy absorbed
W_s = specific energy
W_t = total energy dissipated
α = angle of wedge
δ_{cr} = material parameter

θ = angle of rotation
μ_s = static friction coefficient
μ_d = dynamic friction coefficient
σ = stress
σ_c = tensile fracture stress
$\varphi\ (=\alpha/2)$ = semi-angle of wedge

5.2 GENERAL

In this chapter are reported the small scale modelling of fracture mechanisms and the large scale deformation of fibre-reinforced composite materials when circular thin-walled tubes are subjected to axial collapse [75, 82, 91, 99] and bending [62, 100, 101]. The effect of composite material properties and the tube geometry on the crashworthy characteristics of the tubular structural components is investigated with the aim in mind of the cost-effective design of structural elements in vehicles. A theoretical analysis of the failure mechanism of the stable mode of collapse, based on experimental observations and taking into account all possible energy absorbing mechanisms developed during the process, is also reported. Crushing loads and the energy absorbed are theoretically predicted. The proposed theoretical model was experimentally verified for various composite materials and tube geometries and proved to be very efficient for theoretically predicting the energy absorbing capacity of the shell.

Furthermore, the bending of thin-walled fibreglass composite circular tubes under certain end-clamping conditions, simulating in this manner the oblique collision of structural elements of impacted vehicles, is examined. The fracture mechanism of the crushed composite tubes is quite different than that pertaining to metallic components, loaded under the same conditions. Energy absorption in the present case is achieved by material fragmentation, i.e. the mechanism of fracture dominates the phenomenon rather than the plastic deformation. The effect of clamping conditions on the energy absorbing efficiency of the shell is also examined.

5.3 AXIAL COLLAPSE: STATIC AND DYNAMIC

5.3.1 Experimental

The static and dynamic axial collapse of structural tubular elements fabricated from fibre-reinforced composite material was investigated.

In order to obtain experimental evidence about the governing failure mechanism during the axial loading of thin-walled fibre-reinforced composite circular tubes, static tests were performed between the parallel steel platens of an SMG 100 ton hydraulic press fully equipped and computerised, see Figure 5.1. All tests were carried out at a crosshead speed of 10 mm/min or a compression strain-rate of 10^{-3}/sec.

Dynamic tests were performed by direct impact on a drop-hammer at velocities exceeding 1 m/s. The existing drop- hammer facility, with a 47 kg falling weight (with

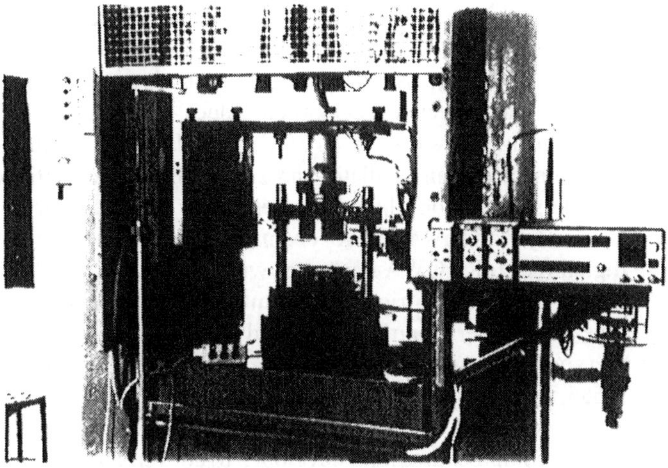

Figure 5.1. A close view of the experimental set-up (hydraulic press) for static axial collapse.

the ability increasing up to 70 kg) from a maximum drop-height of 4 m, provides a maximum impact velocity of about 10 m/s. Load, tube shortening (displacement) and acceleration of the dropping mass during the crushing process were measured and recorded, see Figure 5.2. The specimens rested on a KYOWA strain-gauge load cell of a nominal capacity of 50 ton and accuracy of 5%, connected to a KYOWA signal

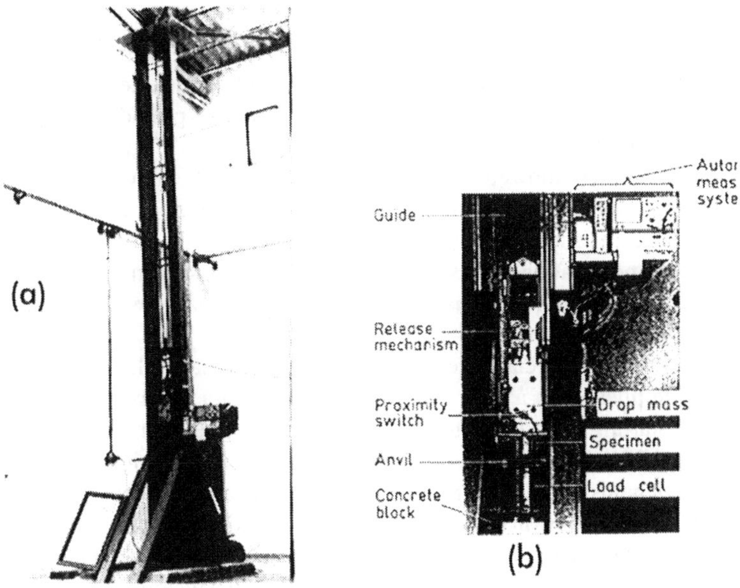

Figure 5.2. (a) A drop-hammer for the dynamic axial collapse. (b) A close view of the experimental set-up and the measuring equipment.

conditioner with a 2500 Hz bandwidth. A KISTLER piezoelectric accelerometer with a measuring range of –50.000g–100.000g was installed on the falling mass and connected to a KISTLER charge amplifier incorporating a low-pass filter of 1 kHz to achieve the levels of the expected acceleration values for the tests performed. A SCHAEVITZ displacement transducer (LVDT) was also used for measuring the tube shortening, connected to an oscillator and a SCHAEVITZ amplifier of a 2.500 Hz bandwidth. The signals from the transducers, after being amplified, were recorded simultaneously via a HAMEG two- channel storage oscilloscope and a 12-bit analogue to digital (A/D) KYOWA converter with accuracy 2.5 mV for a +10 V range connected in parallel and triggered by an inductive proximity switch. The graphs load/time and acceleration/time, recorded on the oscilloscope screen, were also printed by a special thermal printer. The data are also recorded by the converter at a rate of 16.666,7 samples/sec for each of the measured quantities. The data acquired by the A/D converter were transfered by a Grib-IEEE488 to the host HP PC308 computer. A computer program, developed to convert these data to load, acceleration and displacement, respectively, stored and transfered the data to a graphic programme creating the following curves: displacement/time, load/time, acceleration/time, velocity/time (by integration of the acceleration/time curve), load/displacement; see also Figure 5.3.

Two different kinds of fibreglass composite material, designated as composite materials A and B respectively, were used for testing.

Material A is a fibreglass composite material with individual fibre diameter of 9 μm chopped strand mat. The fibre length in the mat was 50 mm with random orientation in the plane of the mat and the glass used was 0.8 gr/mm^2 chopped fibre mat. All specimens were made by a hand lay-up technique by rolling pieces of fibreglass cloth onto a rotating wooden mandrel and, simultaneously, impregnating it with a polyester resin (solution of phthalic anhydride and maleic anhydride esterfied with propylene glycol) using 100 ml of resin per 1000 cm fibreglass and providing in this manner with a composite material of 72% per weight fibre and 1.37 gr/cm^3 density. The specimens were then allowed to cure for 24 h at ambient temperature before they were removed from the mould and then they were finished by polishing, smoothing their exterior seams and squaring their ends. The dimensions of all moulds used for the specimens preparation corresponded to their internal dimensions. The moulds were fabricated from laminated hard maple wooden blocks and were prepared properly before use.

Material B is a commercial fibre reinforced composite material. The tube wall consisted of nine plies, with a total thickness of 2.3 mm for the axial collapse tested specimens. Starting from the exterior of the shell the plies were laid-up in the sequence $[(90/0/2R_c)/(2R_c/0/90)/R_{c.-5}]$, where the 0° direction coincides with the axis of the tube, R_c denotes random chopped strand mat plies and $R_{c.-5}$ represents a similar ply but thinner, providing in this manner with a composite material of 33.9% per volume fibre content and 1.55 g/cm^3 density. The resin formulation is Dow DERAKANE 411-C-50 BZQ-40 1-2%. According to the manufacturer's specifications, the laminate fibre lay-up was hand wrapped around a rigid foam core, whilst plastic staples were used to hold the lay-up on the core during the resin transfer mold-

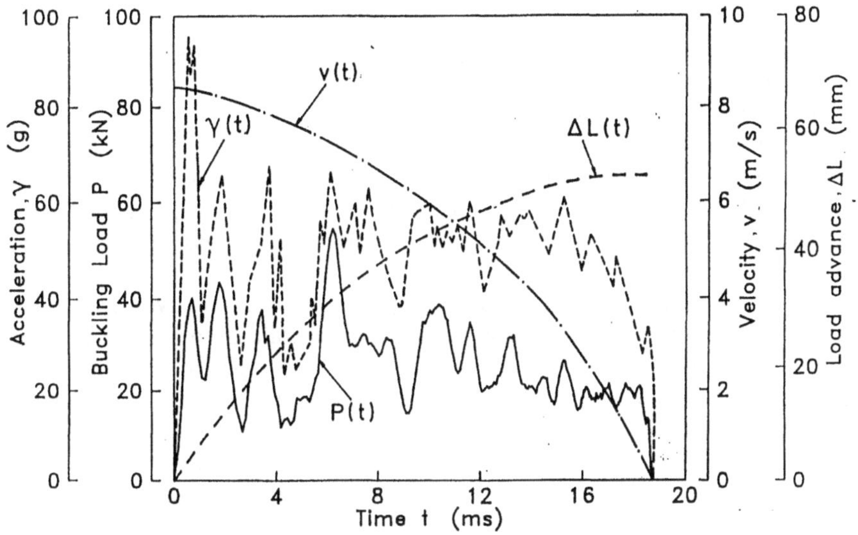

Figure 5.3. (a) Variation of buckling load, displacement, velocity and acceleration with time for dynamically axially collapsed thin-walled composite shells.

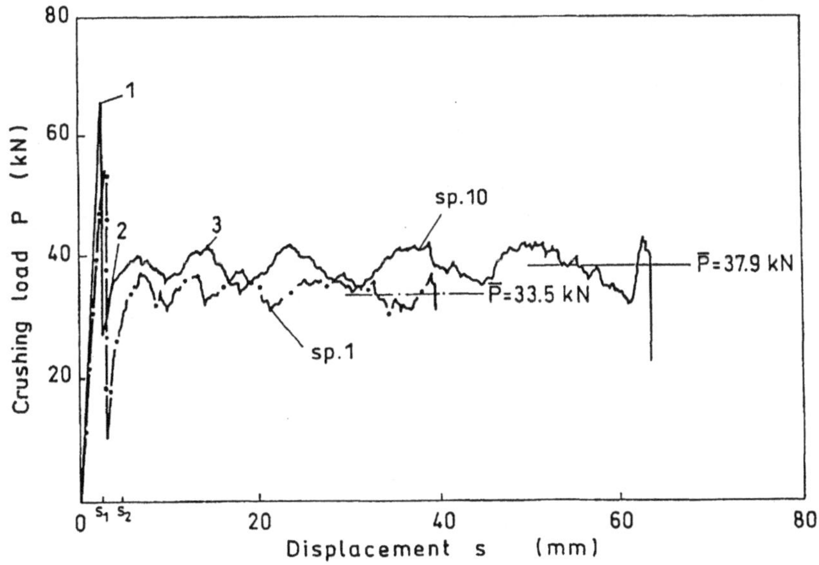

Figure 5.3 (continued). (b) Load/displacement characteristics for tubes of materials A and B (specimens 1 and 10, respectively; see Table 5.1).

ing process. The specimens were injection moulded and allowed to cure for 45 minutes at room temperature and then they were post-cured at 120°C for 3 hours. Finally, the tubes were de-foamed and cut properly.

Details on specimens geometry are given in Table 5.1 for both materials tested, whilst the stress-strain curves, as obtained from quasi-static tension test for both materials A and B, are shown in Figure 5.4.

In order to study the effect of the specimen geometry, care was taken for the specimens to give reproducible results and to absorb acceptable amount of crush energy. Photographs of crushing modes of the specimens tested, along with terminal views and longitudinal sections of the deformed specimens, are shown in Figure 5.5.

To obtain a concept of the failure mechanism, microscopic investigations were performed on crushed circular tubes and frusta, after collapse to a certain amount of deformation, using metallographic techniques. Micrographs of the crush zone at various regions over the tube circumference were obtained by examining metallographic specimens, cut-off from the damaged region of the compressed specimens, on a UNIMET metallographic optical microscope equipped with photographic facilities. To prepare the metallographic specimens, a longitudinal strip at the damaged region was removed from the shell wall and the relevant surface was then prepared and polished properly. Micrographs of the specimens tested are presented in Figures 5.6 and 5.7.

The values of the initial peak load, P_{max} and the total energy absorbed, W_T for the axially loaded specimens, obtained by measuring the area under the load/displacement curve, as well as the mean crushing load, $\overline{P}$, defined as the ratio of energy absorbed to the total shell shortening, and the specific energy, W_S, which is equal to the energy absorbed per unit mass crushed, are tabulated in Tables 5.1(a) and (b) for the static and dynamic tests, respectively.

Table 5.1 Crushing characteristics of axially loaded circular tubes.
(a) Static

Sp. No.	*Mater.*	*Number of layers*	*Thickness, t (mm)*	*Axial Length, L (mm)*	*Mean diam. $\overline{D}$*(mm)*	*$t/\overline{D}$*	*$L/\overline{D}$*	*Crush length, s (mm)*
1	A	3	3.4	102.3	55.0	0.059	1.81	38.7
2	A	4	4.4	107.3	57.0	0.078	1.82	38.6
3	A	5	5.7	106.0	61.2	0.093	1.73	40.4
4	B	**	2.3	127.0	38.0	0.061	3.34	63.5
5	B	**	2.3	101.6	38.0	0.061	2.67	63.5
6	B	**	2.3	76.2	38.0	0.061	2.00	63.5
7	B	**	2.3	50.8	38.0	0.061	1.34	25.4
8	B	**	2.3	25.4	38.0	0.061	0.67	12.7
9	B	**	2.3	12.7	38.0	0.061	0.33	7.0
10	B	**	2.3	101.6	55.6	0.041	1.83	63.5
11	B	**	2.3	76.2	55.6	0.041	1.37	63.5
12	B	**	2.3	25.4	55.6	0.041	0.46	20.0
13	B	**	2.3	12.7	55.6	0.041	0.23	8.5

* $\overline{D} = (D_o + D_i)/2$
** See Section 5.3.1 for material B lay-up

Table 5.1 (continued).

Sp. No.	Collapse Mode	Crushing load, P (kN): Initial peak, P_{max} Exp.	Mean post-crushing, $\bar{P}$: Exp.	Mean post-crushing, $\bar{P}$: Theor.	Total energy absorbed, W_r (kJ): Exp.	Total energy absorbed, W_r (kJ): Theor.	Specific energy, W_s (kJ/kN): Exp.	Specific energy, W_s (kJ/kN): Theor.
1	Ia	54.4	33.5	32.8	1.296	1.296	44.4	45.2
2	Ia	86.8	41.6	43.1	1.607	1.663	44.6	45.3
3	Ia	143.7	62.4	61.0	2.521	2.464	45.8	45.4
4	Ia	31.7	22.3	24.1	1.417	1.530	56.2	57.5
5	Ia	32.5	23.7	24.1	1.509	1.530	56.4	57.5
6	Ia	34.7	24.4	24.1	1.551	1.530	55.5	57.5
7	Ia	31.6	22.9	24.1	0.583	0.612	56.6	57.5
8	Ia	33.2	22.4	24.1	0.184	0.306	57.1	57.5
9	Ia	32.7	22.4	24.1	0.162	0.173	56.7	57.5
10	Ia	65.9	37.9	36.3	2.408	2.305	53.7	54.2
11	Ia	62.6	34.3	36.3	2.175	2.305	53.5	54.2
12	Ia	67.8	40.3	36.3	0.809	0.726	54.1	54.2
13	Ia	64.5	33.4	36.3	0.284	0.308	54.4	54.2

(b) Dynamic

Sp. No.	Mater.	Number of layers	Thickness, t (mm)	Axial Length, L (mm)	Mean diam. $\bar{D}$*(mm)	Crush speed, v (m/s)	Crush length, s (mm)
14	A	3	3.1	97.7	56.0	6.0	33.7
15	A	4	4.5	105.4	58.8	6.0	18.6
16	A	5	5.7	105.5	64.1	6.0	15.2
17	A	3	3.0	100.1	95.8	7.0	38.6
18	A	4	4.6	104.8	59.0	7.0	23.4
19	A	5	5.6	103.8	64.0	7.0	17.4
20	A	3	3.1	100.9	56.1	8.1	51.8
21	A	4	4.5	105.5	58.8	8.1	31.1
22	A	5	5.8	105.6	64.4	8.1	24.8
23	B	**	2.3	127.8	38.0	6.7	33.7
24	B	**	2.3	101.5	38.0	6.7	33.2
25	B	**	2.3	76.6	38.0	6.7	37.1
26	B	**	2.3	128.2	55.6	6.7	24.0
27	B	**	2.3	100.0	55.6	6.7	25.6
28	B	**	2.3	75.2	55.6	6.7	25.8

* $\bar{D} = (D_o + D_i)/2$
** See Section 5.3.1 for material B lay-up

Table 5.1 (continued).

Sp. No.	Collapse Mode	Crushing load, P (kN): Initial peak, P_{max} Exp.	Mean post-crushing, $\overline{P}$ Exp.	Mean post-crushing, $\overline{P}$ Theor.	Total energy absorbed, W_r (kJ) Exp.	Total energy absorbed, W_r (kJ) Theor.	Specific energy, W_s (kJ/kN) Exp.	Specific energy, W_s (kJ/kN) Theor.
14	Ia	40.1	23.9	22.8	0.900	0.873	34.7	33.6
15	Ia	96.8	43.6	41.8	0.812	0.800	41.9	38.8
16	Ia	64.8	50.7	52.2	0.774	0.796	35.8	37.2
17	Ia	39.6	28.0	27.7	1.086	1.012	41.9	39.3
18	Ia	61.8	42.7	41.9	0.998	0.904	39.6	39.5
19	Ia	64.1	50.5	51.6	0.884	0.910	36.0	38.4
20	Ia	40.1	25.7	23.2	1.333	1.202	36.4	32.8
21	Ia	44.4	41.7	42.5	1.297	1.322	39.7	40.5
22	Ia	96.9	50.4	47.1	1.250	1.168	38.3	35.8
23	Ia	30.7	29.5	32.2	0.994	1.085	66.5	72.6
24	Ia	33.4	29.4	32.2	0.977	1.069	66.4	72.4
25	Ia	30.9	27.9	32.2	1.035	1.195	62.9	72.6
26	Ia	56.2	41.2	40.8	0.990	0.979	62.3	61.6
27	Ia	51.3	39.9	40.8	1.021	1.044	60.2	61.6
28	Ia	57.0	36.7	40.8	0.948	1.053	62.1	61.6

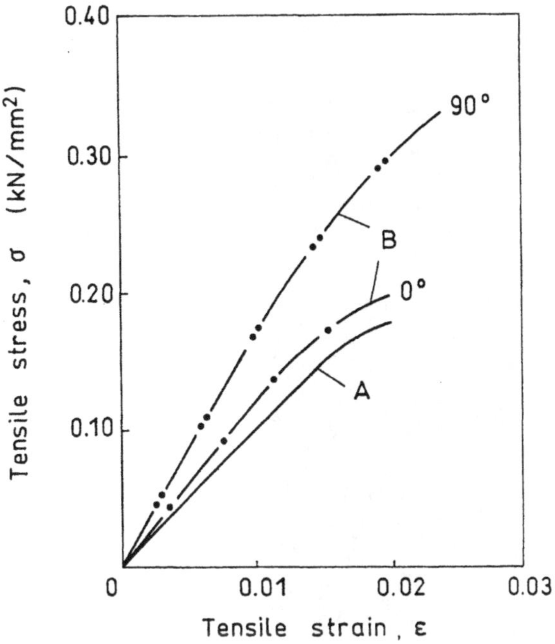

Figure 5.4. Stress/strain curves of composite materials A and B.

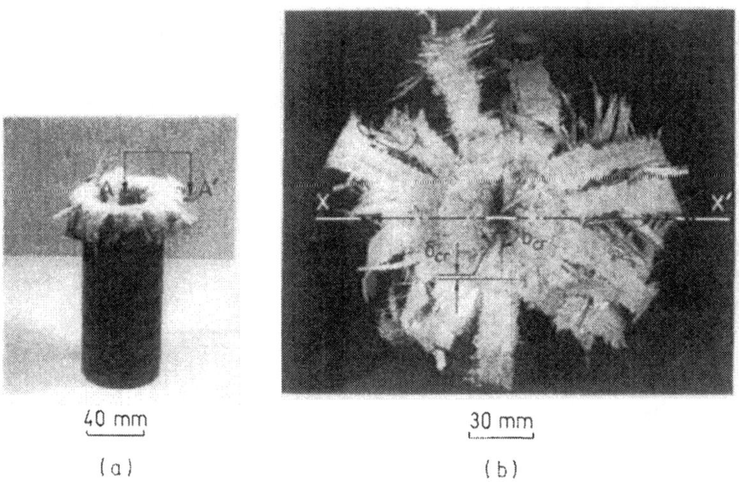

Figure 5.5. Terminal crushing modes for specimen 10 (see Table 5.1): (a) side view; (b) top view.

(a)

(b)

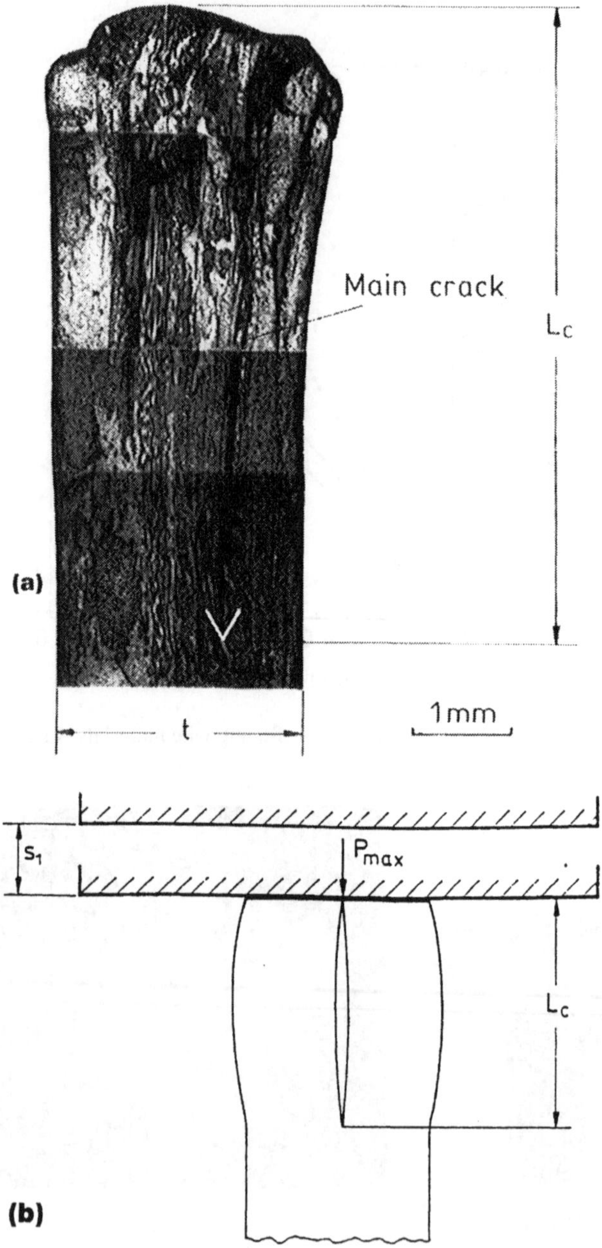

Figure 5.6. (a) Micrograph at section AA′ of Figure 5.5(a) showing microfailures of the crush-zone corresponding to point 1 of the load-displacement curve of Figure 5.3(b) (sp. 10). (b) Configuration of internal fracture mechanism in the crush zone of (a).

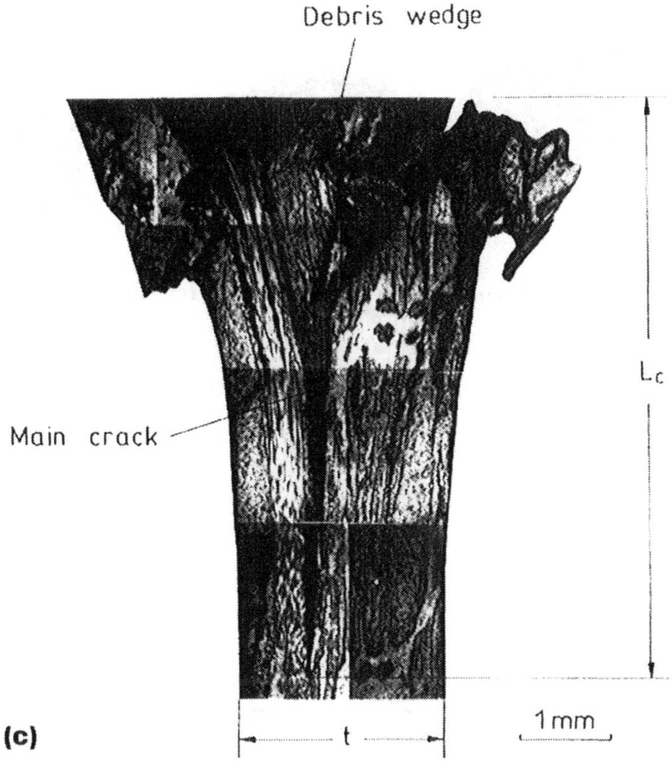

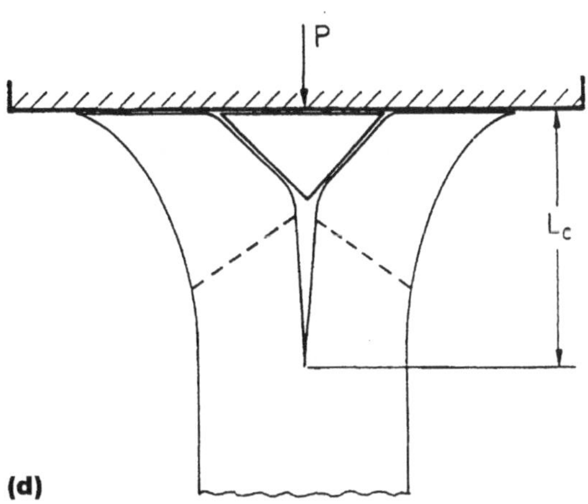

Figure 5.6 (continued). (c) Micrograph at section AA′ of Figure 5.5(a) showing micro-failures of the crush-zone corresponding to point 2 of the load-displacement curve of Figure 5.3(b) (sp. 10). (d) Configuration of internal fracture mechanism in the crush zone of (c).

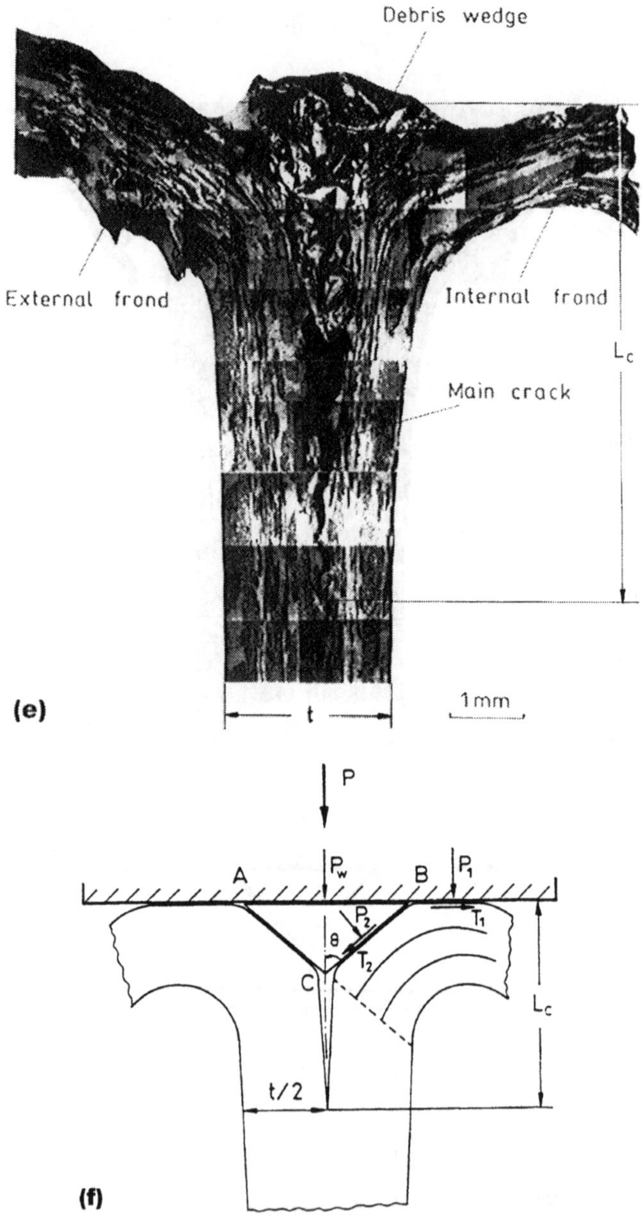

Figure 5.6 (continued). (e) Micrograph at section AA′ of Figure 5.5(a) showing microfailures of the crush-zone corresponding to point 3 of the load displacement curve of Figure 5.3(b) (sp. 10). (f) Configuration of internal fracture mechanism in the crush zone of (e).

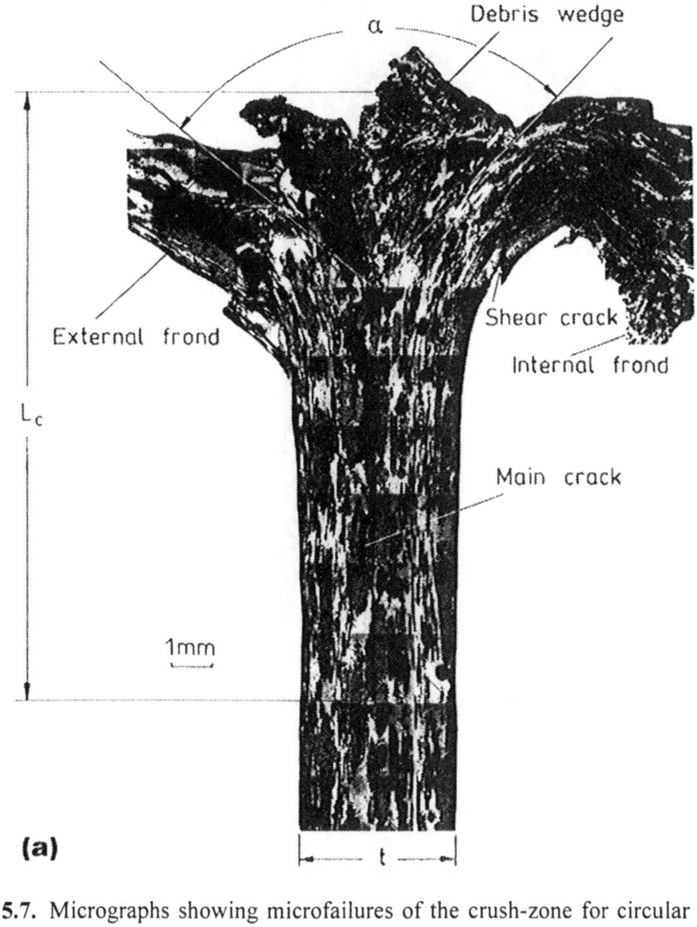

Figure 5.7. Micrographs showing microfailures of the crush-zone for circular tubes (a) statically loaded (sp. 11; see Table 5.1).

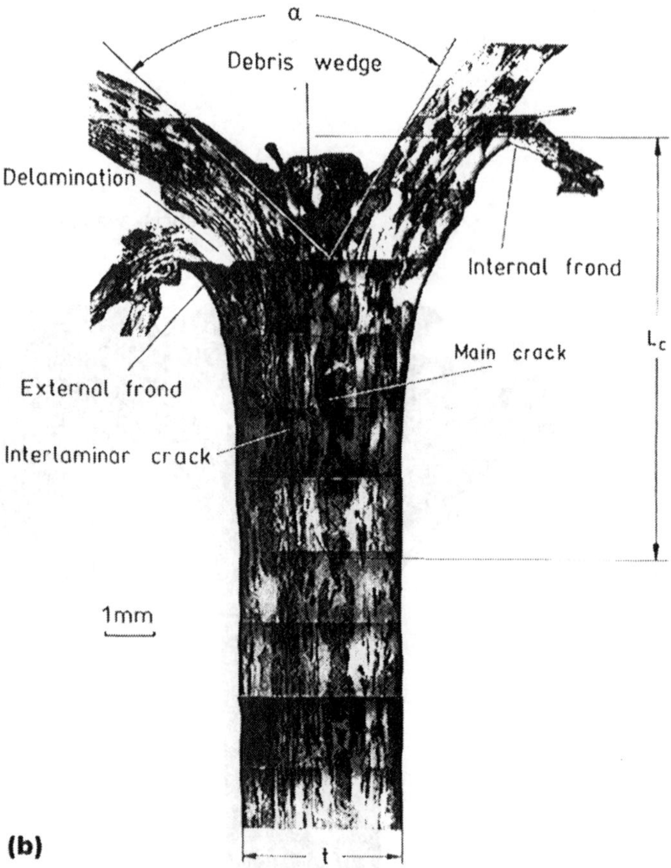

(b)

Figure 5.7 (continued). Micrographs showing microfailures of the crush-zone for circular tubes (b) dynamically loaded (sp. 25; see Table 5.1).

5.3.2 Failure Mechanisms: Experimental Observations

The specimens tested collapsed following the failure Mode I, associated with the progressive collapse of the shell and the formation of continuous fronds, which spread radially outwards and inwards in the form of a "mushrooming" failure, see Figure 5.5 and Table 5.1. Micrographs at various regions of the crush zone and for different deformations over the tube circumference were obtained by examining metallographic specimens, cut-off from the damaged region of the compressed specimens.

As deformation proceeds, the externally formed fronds curl downwards with the simultaneous development of a number of axial splits, see Figure 5.5 (b), due to the developed tension in the circumferential direction of the shell, followed by splaying of material strips. They mainly depend on shell material and stacking conditions. The length of the splits probably derive the effective column length of the material strips undergoing loading. It was clearly observed that the formation of these strips was more distinct and the number of the splits was greater for the specimens of material B. Axial tears were not apparent in the internal fronds which were more continuous than their external counterparts for both materials.

Initially the shell behaves elastically and the load rises at a steady rate to a peak value, P_{max} and then drops abruptly. At this stage, a central intrawall crack of length L_c forms at the end of the shell adjacent to the loading area, see the experimentally obtained micrograph of the crush region in Figure 5.6(a) and the schematic diagram in Figure 5.6(b). Note, however, that the formation of additional longitudinal interply cracks, at both sides of the main central one, see Figure 5.6(a), may be attributed to bending due to non-parallel alignment between the tube and the press platen. Due to the elastic deformation of the shell a bulge of the upper edge of the shell adjacent to the loading area is formed, see also Figure 5.6(b).

In the post-crushing regime the load increases with increasing deflection and then it starts oscillating about a mean post-crushing load, $\overline{P}$; the formation of the first peak and drop of the load, see Figure 5.3(b), may be associated with the formation of two equal lamina bundles, bent inwards and outwards, due to the flexural damage, which occurs after the abrupt load drop at a distance from the contact surface equal to the wall thickness, t. At this stage, a triangular debris wedge of pulverised material starts to form; its formation may be attributed to the friction between the bent bundles and the platen, see Figures 5.6(c) and (d). Note that the wedge is completed when the crushing load starts oscillating and subsequently its size remains constant throughout the deformation process in the post-crushing regime, see Figures 5.6(e) and (f). As measured from the specimen tested, the wedge angle, α (=2φ), is 100°–110°.

The behaviour of the reinforcing fibres depends upon their orientation. Axially aligned fibres bend inwards or outwards, with or without fracturing, according to their flexibility and the constraints induced by other fibres; their effective flexibility depends upon the fibres structure in the composite material. Fibres aligned in the hoop direction can only expand outwards by fracturing and inwards by either fracturing or buckling.

Delamination occurs as a result of shear and tensile separation between plies. The axial laminae split into progressively thinner layers, forming, therefore, translaminar cracks normal to the fibres direction, mainly due to fibre buckling, finally resulting either in fibre fracture or in intralaminar shear cracking which splits the laminate into many thin layers without fibres fracture. Cracks propagate preferably through the weakest regions of the structure of the composite material, i.e. through resin-rich regions or boundaries between hoop fibres, resulting in their debonding, or through the interface between hoop and axial plies causing delamination.

The following principal sources of energy dissipation at microscopic scale, contributing to the overall energy absorption during collapse, may be listed:

- Intrawall crack propagation
- Fronds bending owing to delamination between plies

- Axial splitting between fronds
- Flexural damage of individual plies due to small radius of curvature at the delamination limits
- Frictional resistance to axial sliding between adjacent laminae
- Frictional resistance to the penetration of the annular debris wedge
- Frictional resistance to fronds sliding across the platen

The values of the initial peak load, P_{max}, the energy absorbed, W (obtained by measuring the area under the load/displacement curve) and the mean crushing load, $\overline{P}$, defined as the ratio of energy absorbed to the total shell shortening, are tabulated in Table 5.1.

In general, the microfracture mechanism of Mode Ia of collapse is similar for statically and dynamically loaded shells. The only differences encountered are associated with the shape of the pulverised wedge and the microcracking development.

The size of the debris wedge and the main central crack dimensions are smaller in the case of the impacted shells; compare Figures 5.7(a) and (b). During dynamic loading, the resin behaved in a more brittle manner and it was shattered and separated almost completely from the fibres in the crush zone, an indication of the brittle behaviour of the composite material exhibited during its loading at elevated strain-rates. The wedge angle, α was estimated to about 90°, whilst in the case of the statically loaded specimens it was about 100°–110°.

5.3.3 Energy Absorbing Characteristics

Typical load/displacement curves for static and dynamic loading of both materials A and B, i.e. the variation of crushing load, P with shortening of the shell for the end-crushing Mode I, are shown in Fig 5.3(b). During the elastic regime, the load rises at a steady rate to a peak value, P_{max}, which depends on shell geometry and material characteristics, and then drops abruptly. As mentioned above for statically loaded tubes, in the post-crushing regime the load increases with increasing deflection and then it starts oscillating about a mean post-crushing load, $\overline{P}$.

Dynamically obtained load/displacement curves show more severe fluctuations, with troughs and peaks, than the corresponding statically ones, see also Figure 5.3(b). From the shape of the dynamically obtained curves in the post-crushing region, it was rendered difficulties to assuming the possible fracture mechanism occurred, as well as the development and propagation of the microcracks. On the contrary, the deformation mode encountered during static collapse could be excluded from the pattern of a load/displacement curve (history of deformation). Moreover, due to the dynamic nature of the phenomenon (note that its duration ranges between 7–25 ms), the sequence of different microcracking processes in relation to the shape of the obtained load/displacement curve can not be followed distinctly.

Taking into account the observations reported above, two different fracture mechanisms may be proposed to explaining the shape of curves for dynamically loaded specimens. They are based on observations related to crack propagation in composite materials, neglecting friction forces and bending, as well as microscopic observations regarding the collapse mode of shells. However, the dynamic nature of the process must not be overlooked and, therefore, phenomena, such as oscillations associated with the inertia problems, may be also taken into account:

- According to the first mechanism, cracking develops simultaneously with friction and bending, however, it propagates faster than the associated mechanical phenomena.
- Cracks and delaminations develop abruptly after a certain amount of deformation is imposed, followed by friction and bending; this is repeated successively until the final collapse of the shell.

5.3.4 Failure Analysis

STATIC AXIAL COLLAPSE

Consider the failure mechanism, Mode I, described above in detail. During the elastic deformation of the shell the load rises at a steady rate to a peak value, P_{max}. At this stage, a central intrawall crack of length L_c forms at the top end, see the micrograph of Figure 5.6 (a) and the schematic diagram in Figure 5.6 (b); the related tube shortening is s_1, see Figure 5.3(b).

If R_{ad} is the fracture energy required to fracture a unit area of the adhesive at the interface between two adjacent layers, the energy dissipated for the intrawall crack to propagate at length L_c can be calculated by applying fracture theory, see Reference [46] and also Section 3.3.5 in Chapter 3. Note that this energy equals the external work, as can be obtained by measuring the area under the load/displacement curve in the elastic regime, see Figure 5.3(b). Therefore,

$$W_{Lc} = R_{ad} \cdot (\pi \cdot \overline{D} \cdot L_c) = \int_0^{s_1} P ds = \frac{1}{2} P_{max} s_1 \tag{5.1}$$

where $\overline{D} = (D_o + D_i)/2$, is the mean diameter of the tube; D_o and D_i are the outside and inside diameter, respectively.

As mentioned above, with deformation progressing in the post-crushing regime, the load increases with increasing deflection associated with the formation of two equal lamina bundles bent inwards and outwards. The triangular debris wedge of pulverised material starts to form over the shell circumference and is completed when the crushing load starts oscillating, with its size remaining constant throughout the deformation process in the post-crushing region; the related tube shortening, corresponding to the completion of the wedge formation, is s_2.

The energy required for this deformation mechanism regarding the formation of the crush zones, see Figures 5.6 (c) and (d), equals the external work absorbed by the deforming shell in this regime, i.e.

$$W_{tr} = [2 \int_0^{\varphi} \sigma_o \cdot l_s \cdot (l_s/2) d\varphi] \cdot \pi \cdot \overline{D} = \int_{s_1}^{s_2} P ds \tag{5.2}$$

where, following the Notation, σ_o is the normal stress applied by the wedge to fronds, l_s the side length of the wedge inscribed to the bent fronds and φ (= $\alpha/2$) the semi-angle of the wedge.

It is also assumed that the intrawall crack propagates with a constant speed equal to the speed of the crosshead of the press and, therefore, its lentgth L_c remains constant. Also, the length of the split of the crush zone (AB) at the contact side with the steel platen approximates the wall thickness, t, see Figures 5.6(e) and (f).

As loading proceeds further, i.e. after the displacement s_2, resulting in crushing with the subsequent formation of the internal and external fronds, normal stresses develop on the sides of the debris wedge followed by shear stresses along the same sides due to the friction at the interface between the wedge and the fronds. Note, also, that additional normal and shear stresses develop at the interface between the steel press platen and the deforming shell as the formed fronds slide along this interface.

Static equilibrium at the interface yields

$$P = (P_w + 2 \cdot P_1) \cdot \pi \cdot \overline{D} \tag{5.3}$$

where, P_w is the normal force per unit length applied by the platen to the debris wedge and P_1 the normal force per unit length applied by the platen to the internal and external fronds.

The normal force applied by the platen to the debris is equilibrated by normal and frictional components at the uppermost frond surfaces, see Figure 5.6 (f); therefore,

$$P_w = 2 \cdot (P_2 \cdot \sin\varphi + T_2 \cdot \cos\varphi) \tag{5.4}$$

where, P_2 is the normal force per unit length applied to the sides of the wedge and T_2 the frictional force per unit length developed between wedge and fronds.

Assuming that Coulomb friction prevails between the debris wedge and fronds, $T_2 = \mu_{s2} \cdot P_2$, Eq. (5.4) becomes

$$P_w = 2 \cdot P_2 \cdot (\sin\varphi + \mu_{s2} \cdot \cos\varphi) \tag{5.5}$$

where, μ_{s2} is the coefficient of friction.

It must be noted that

$$P_2 = \sigma_o \cdot l_s \tag{5.6}$$

and

$$\sigma_o = k \cdot \sigma_\theta \tag{5.7}$$

where, k is a constant and σ_θ the tensile fracture stress of the composite material.

From the shape of the wedge is

$$l_s = t/(2 \cdot \sin\varphi) \tag{5.8}$$

Substituting Equations (5.6)–(5.8) into Equation (5.5), P_w is determined and hence P_1.

The horizontal component of the force F_x applied by the wedge to the frond is

$$F_x = P_2 \cdot (\cos\varphi - \mu_{s2} \cdot \sin\varphi) \tag{5.9}$$

Taking into account that $F_x > 0$

$$\varphi < \tan^{-1}(1/\mu_{s2}) \tag{5.10}$$

For μ_{s2} less than unity, φ can not be smaller than 45°. Note, also, that the maximum value of φ depends on μ_{s2}.

The energy dissipated in frictional resistance for a crush distance s is

$$W_i = 2 \cdot (\mu_{s1} \cdot P_1 + \mu_{s2} \cdot P_2) \cdot \pi \cdot \overline{D} \cdot s \tag{5.11}$$

and the corresponding load

$$P_i = 2 \cdot (\mu_{s1} \cdot P_1 + \mu_{s2} \cdot P_2) \cdot \pi \cdot \overline{D} \tag{5.12}$$

where, μ_{s1} is the coefficient of friction between frond and platen, mainly dependent on the platen material and the surface finish.

Energy is also dissipated due to bending of the fronds, see Figures 5.6 (e) and (f), and, therefore, for a subsequent displacement s, it results in

$$W_{ii} = 2 \cdot [\int_0^{\varphi} P_2 \cdot (l_s/2) d\varphi + \int_{s_2}^{s} P_2 \cdot \varphi \cdot ds] \cdot \pi \cdot \overline{D} \tag{5.13}$$

The related crushing load is

$$P_{ii} = W_{ii}/s \tag{5.14}$$

The main central intrawall crack is assumed to continue to propagate at a constant speed equal to the crosshead of the press and, therefore, for a crush displacement s the energy dissipated is

$$W_{iii} = R_{ad} \cdot [(s - s_1) + L_c] \cdot \pi \cdot \overline{D} \tag{5.15}$$

and the corresponding load

$$P_{iii} = W_{iii} / s \tag{5.16}$$

The energy absorbed by the external fronds for the axial splits formed due to developed tensile stresses in the hoop direction of the shell, as these fronds curl downward, and the subsequent splaying of material strips, see Figure 5.4(b), is

$$W_{iv} = n \cdot (t/2) \cdot G \cdot s \tag{5.17}$$

where, n is the number of splits and G the fracture toughness.

The corresponding force required is

$$P_{iv} = n \cdot (t/2) \cdot G \tag{5.18}$$

The number of axial splits n can be determined from the ratio

$$n = 2 \cdot \pi / b_{cr} \tag{5.19}$$

when

$$b_{cr} = 2 \cdot \cos^{-1} [1 - \delta_{cr} / ((D_i/2) + t)] \tag{5.20}$$

where, b_{cr} is the angle between two splits and δ_{cr} the critical distance at which axial splitting occurs; see Figure 5.4(b).

From Equations (5.11), (5.13), (5.15) and (5.17) the total energy dissipated for the deformation of the shell is

$$\begin{aligned} W_T = {} & [1/(1 - \mu_{s1} + \mu_{s1} \cdot s_2 / s] \cdot [\pi \cdot (d - t) \cdot t \cdot k \cdot \sigma_\theta \cdot \{(s - s_2) \cdot [\mu_{s2} / \cos(\alpha/2) \\ & - \mu_{s1} \cdot [\tan(\alpha/2) + \mu_{s2}]] + [(\alpha/2)/\cos(a/2)] \cdot [0.25 \cdot t / \cos(a/2) \\ & + s - s_2]\} + R_{ad} \cdot \pi \cdot (d - t) \cdot (s - s_1 + L_c) + n \cdot (t/2) \cdot G \cdot s] \end{aligned} \tag{5.21}$$

where, α is the angle of the pulverised wedge.

The total normal force applied by the platen to the tube can be calculated as

$$P = W_T / s \tag{5.22}$$

From the stress-strain curves shown in Figure 5.4, σ_θ of the materials was estimated. The static friction coefficients, μ_{s1} and μ_{s2}, were obtained by employing the curling test [41]. The fracture toughness, G was estimated from the tension test of notched strips. The interfacial fracture energy, R_{ad} was calculated from Equation (5.2), using the experimental results obtained by loading

Table 5.2 Material properties.

Material	R_{ad} (kJ/mm^2)	G (kJ/mm^2)	σ_θ (kN/mm^2)	μ_{s1}	μ_{s2}	μ_{d1}	μ_{d2}	k
A	0.011	0.020	0.18	0.30	0.55	0.28	0.50	0.14
B	0.050	0.192	0.33	0.33	0.62	0.36	0.70	0.07

cylindrical tubes up to the maximum load, P_{max} and obtaining the energy absorbed from the related experimental load-displacement curves of loaded shells. Note, however, that the R_{ad} may be also estimated by employing the strip peel test, see Reference [46].

Experimentally obtained values for μ_{s1}, μ_{s2}, R_{ad}, G, σ_θ and the constant k are presented in Table 5.2.

THE EFFECT OF STRAIN-RATE

As mentioned above, the microfracrure mechanism for the progressive collapse of axially impacted circular tubes is similar to that of statically loaded ones of the same geometry. However, experimental observations revealed that the size of the debris wedge, as well as the dimensions of the main central crack developed, are smaller in the case of the tubes subjected to dynamic loading; compare Figures 5.6(a) and (b).

The effect of strain-rate in the crushing behaviour of axially loaded composite tubes is reviewed in References [66, 84]. Strain-rate and, therefore, crush-speed may influence the mechanical properties of the fibre and the matrix of the composite material. Brittle fibres are, in general, insensitive to strain-rate, whilst the fracturing of the lamina bundles is not a function of crush-speed. Strain-rate has an effect on matrix stiffness and failure strain and, therefore, the energy absorption, associated with the interlaminar crack growth, may be considered as a function of the crush-speed [66]. Note that the coefficient of friction, between the composite material and the crushing surface and between the debris wedge and the fronds, and, subsequently, the energy absorption capability of the shell are also affected from the crush-speed [66, 77].

The major part of the absorbed energy during the static axial compression of a shell is dissipated as frictional work in the crushed material, or at the interface between the material and the tool, and this is estimated to be about 50% or more of the total work done; see References [75, 77, 93]. In dynamic collapse, attention is directed towards the influence of strain-rate on the frictional work absorbed during the impact, taking into account all structural and material parameters which may contribute to it, i.e. the fibre and matrix material, the fibre diameter and orientation in the laminate, the fibre volume content as well as the conditions at the interfaces.

The total energy absorbed during the dynamic axial collapse of circular tubes may be, therefore, estimated from Equation (5.21) by substituting only the static friction coefficients μ_{s1}, and μ_{s2} with the dynamic ones μ_{d1}, and μ_{d2}, respectively, as

$$W_T = [1/(1-\mu_{d1} + \mu_{d1} \cdot s_2/s)] \cdot [\pi \cdot (d-t) \cdot t \cdot k \cdot \sigma_\theta \cdot \{(s-s_2) \cdot [\mu_{d2} / \cos(\alpha/2) - \mu_{d1} \cdot [\tan(\alpha/2) + \mu_{d2}]] + [(\alpha/2)/\cos(\alpha/2)] \cdot [025 \cdot t/\cos(\alpha/2) + s - s_2]\} + R_{ad} \cdot \pi \cdot (d-t) \cdot (s - s_1 + L_c) + n \cdot (t/2) \cdot G \cdot s] \tag{5.23}$$

An estimate of the dynamic coefficients of friction at the fronds/platen interface, μ_{d1} and the fronds/wedge intertace, μ_{d2} was obtained by comparing the experimental crushing loads in Table 5.1. The estimated values are presented in Table 5.2.

5.3.5 Crashworthy Capability: Concluding Remarks

The axial static and dynamic loading of fibre-reinforced composite circular tubes was experimentally investigated. The failure modes, obtained during the present series of static and impact collapse of circular tubes made of material A and B, were only stable collapse modes (Mode I) and are also encountered, see Table 5.1. The microfracture mechanism, pertaining to the progressive mode of collapse observed, was also theoretically analysed.

Even though the transitional crushing phenomena, the formation and the growth of the collapse mechanism concerning the axial collapse of circular tubes of fibreglass composite materials are complicated and stochastic, the described failure mechanism was verified both theoretically and experimentally. In Table 5.1 experimental measurements and theoretical predictions of crushing load and energy dissipated are listed.

The mean post-crushing load, $\overline{P}$, the total energy absorbed, W_T, and the specific energy, W_s are well predicted theoretically by the proposed analysis within ±10%, see Table 5.1. Note that the proposed theoretical approach is valid for all fibreglass composite materials, provided that the geometry and materials properties are known, but it gives better results for shells with even number of layers, see also Table 5.1. However, in this approach fibre orientation is not taken into account. Also, from Table 5.1 it is evident that the mean post-crushing load and the energy absorbed are mainly affected by the crushing length, whilst the axial length of the tube has no significant effect on the mechanical response of the shell due to failure and, therefore, on these crashworthy characteristics. For the prediction of the crushing loads and the absorbed energy of the dynamically loaded shells, it was assumed that the tensile fracture stress and the fracture toughness of the material remain constant, whilst these parameters can be a function of the crush-speed.

The energy distribution of crashworthy phenomena, pertaining to the axial collapse of composite multi-layered shells, is governed by various parameters; in the proposed simplified theoretical approach the following parameters were encountered:

- Friction between annular wedge and fronds and between fronds and platen
- Fronds bending
- Crack propagation
- Axial splitting between fronds

From the proposed analysis, the distribution of the dissipated energy of the crushed shell, associated with the main energy sources, can be estimated as: Energy absorbed, due to friction between the annular wedge and the fronds and between the fronds and the platen about 50%, due to bending of the fronds about 40%, due to crack propagation about 7% and due to axial splitting about 3%. It is, therefore, evident that the frictional conditions between wedge/fronds and fronds/platen constitute the most significant factors to the energy absorbing capability of the shell; the friction coefficients, μ_{s1}, μ_{d1} and μ_{s2}, μ_{d2} are greatly affected by the surface conditions at the interfaces between composite material/platen or drop-mass and composite material/debris wedge, respectively.

For tubes made of composite material A and subjected to dynamic loading, lower values of the energy absorbed, by about 20%, were obtained, as compared with those predicted in static collapse, probably due to the lower values of the dynamic friction coefficients between the wedge/fronds and fronds/platen interfaces and to the effect of the strain-rate on the microfracture mechanism, outlined above. Contrariwise, in the case of circular tubes made of composite material B, dynamic collapse overestimates static collapse by about 15%, probably due to the different values of the dynamic coefficients for material B, as compared to the related ones for material A, see Table 5.2. It seems that the governing mechanism in this case is the friction mechanism rather than the changes in microfailures. Therefore, it may be concluded that the characteristics of the shell material greatly affect the crashworthy behaviour of impact loaded shells.

It was also observed that the energy absorbed due to bending of the fronds is greatly affected by the magnitude of the wedge's semi-angle, which varies between 45° and $\varphi = \tan^{-1}(1/\mu_2)$.

The number of axial splits, which is affected by the material properties, the lay-up and shell geometry, is greater for the tubes of material B. In general, from the energy absorbing capacity point of view, circular tubes of material B seem to be more efficient, compared to circular tubes of material A, as predicted both theoretically and experimentally, see Table 5.1; this is mainly due to the fabrication and the increased strength of the former material, see Figure 5.4.

5.4 BENDING

5.4.1 Experimental

The experimental set-up used is shown in Figure 5.8. The specimen was clamped at one end suitably and supported at a point close to its opposite end. The torque required for the tube bending was supplied by a speed reducer driven by an electric motor. The torsional load cell and the tube holding fixture were attached to the output shaft of the speed reducer. The load cell was connected to an Ellis bridge amplifier, with the output connected directly to the y-axis of an x-y plotter, and was calibrated by hanging known weights on a specimen placed on a platform

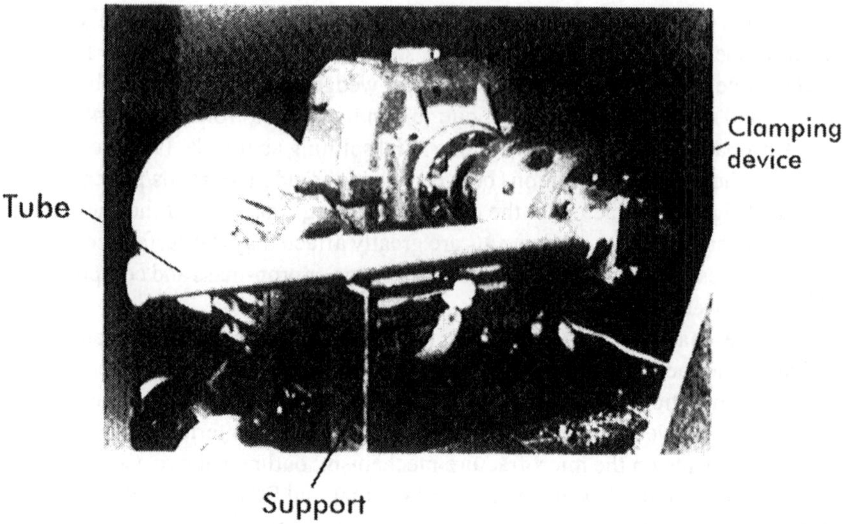

Figure 5.8. Experimental set-up for bending.

suspended from the specimen at a distance of 250 mm from the centre-line of the load cell. The system was zeroed and the weights were placed on the platform producing a torque on the specimen. The x-y plotter was then adjusted to an appropriate torque scale.

The measurement of the angle of rotation was attained by modifying the end of the speed reducer and attaching a potentiometer to the opposite end of the load cell shaft, which was used to convert the angle of rotation to a voltage and then to drive the x-axis of the x-y plotter. The system of angle measurement was calibrated by rotating the fixture from 0° to 90°. A level was placed on the top of the tube fixture and the system was zeroed. The speed reducer was then rotated until the next edge of the tube fixture (machined at a 90° angle) was closed to the top position. The level was then placed on this face and rotation continued until the level indicated 90. Axis adjustment was made using the gain of the *x-y* plotter to give a convenient scale reading. Each specimen was then loaded by turning the speed reducer very slowly through an angle θ, bending in this manner correspondingly the tube through an equal angle. In order to investigate the influence of clamping conditions on the bending moment characteristics, six different holding fixtures were used; see the classification in (a) of Figures 5.9–5.14.

Commercially fibre-reinforced composite tubes, designated as material B, were used as test specimens; see Section 5.3.1 for details about the fabrication and the lay-up of material B. Details related to the specimen dimensions and testing conditions are presented in Table 5.3.

Longitudinal strips of the tube wall from the various crushed regions were cut and encapsulated in potting resin and the relevant surfaces were then smoothed on 200,

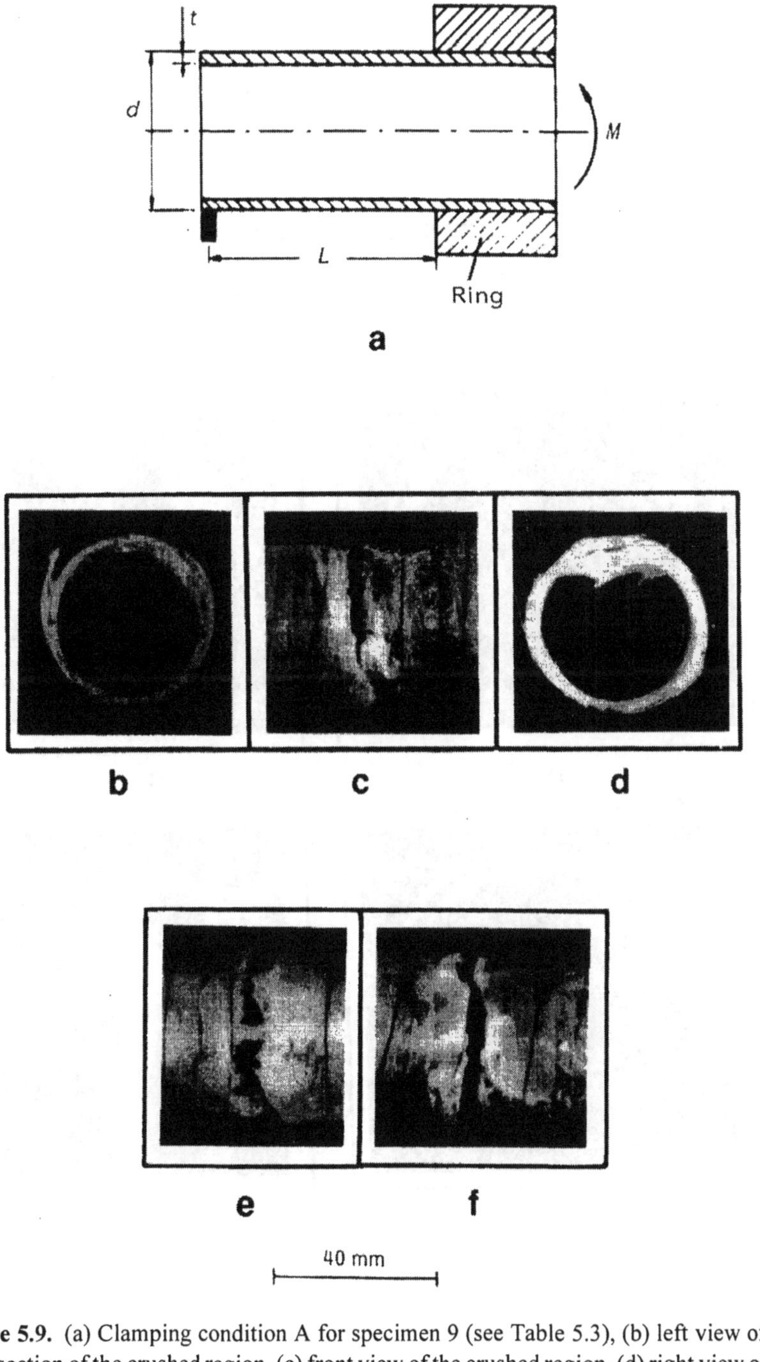

Figure 5.9. (a) Clamping condition A for specimen 9 (see Table 5.3), (b) left view of the cross section of the crushed region, (c) front view of the crushed region, (d) right view of the cross section of the crushed region, (e) tension zone, (f) compression zone.

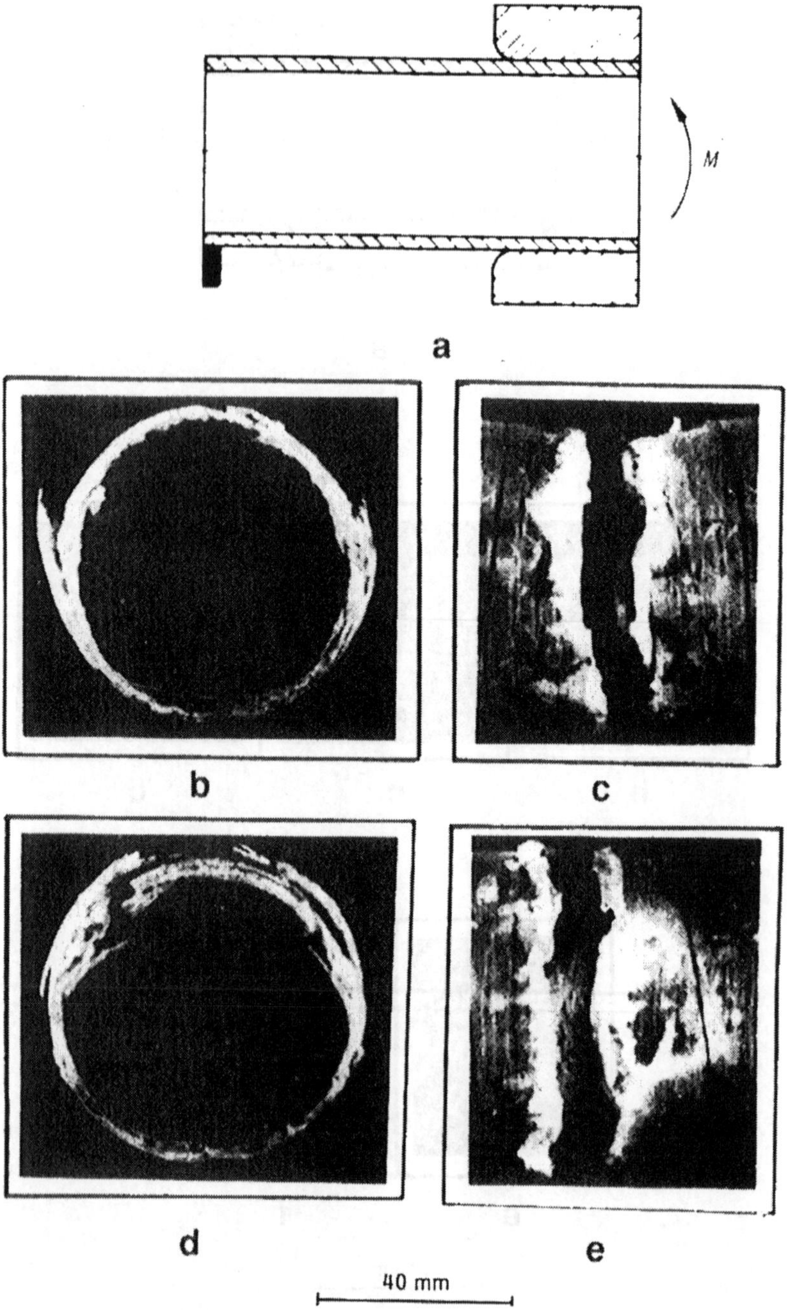

Figure 5.10. (a) Clamping condition B for specimen 5 (see Table 5.3), (b) left view of the cross section of the crushed region, (c) front view of the crushed region, (d) right view of the cross section of the crushed region, (e) compression zone.

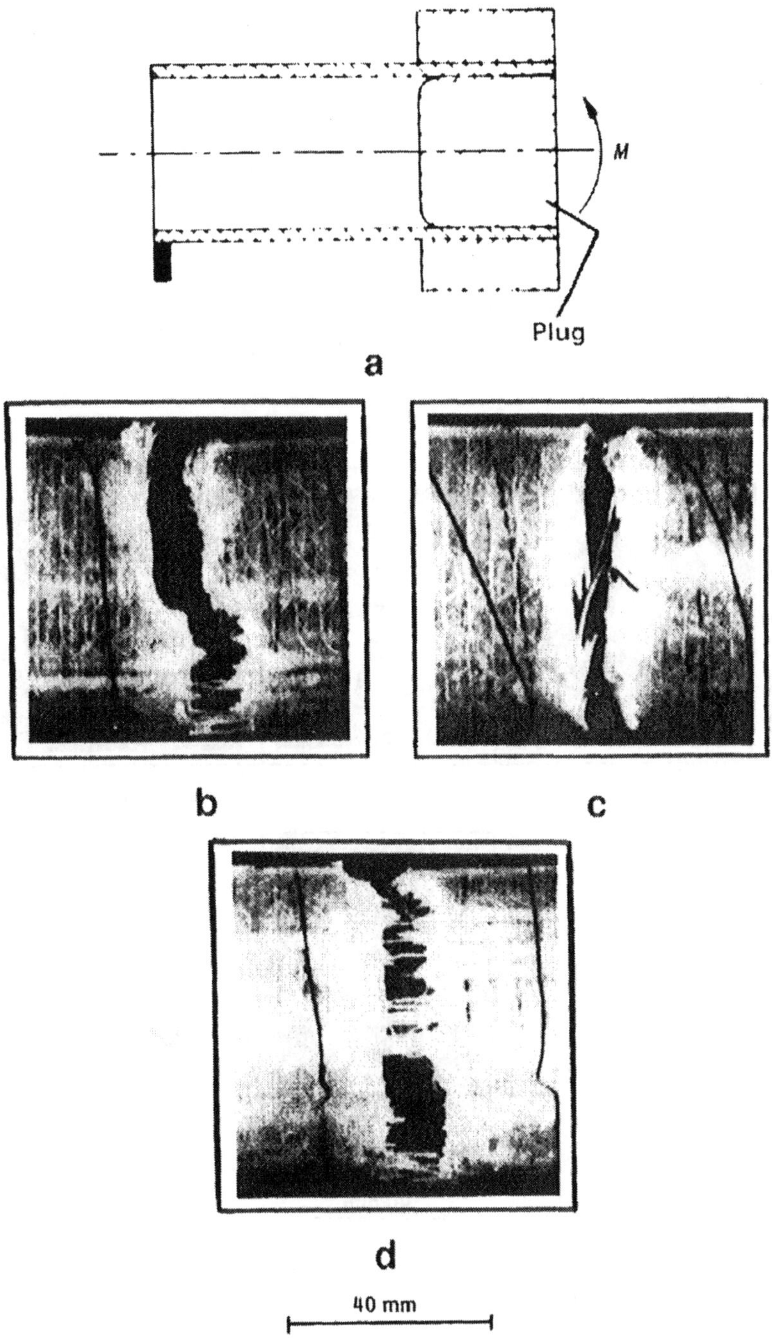

Figure 5.11. (a) Clamping condition C for specimen 2 (see Table 5.3), (b) front view of the crushed region, (c) compression zone, (d) tension zone.

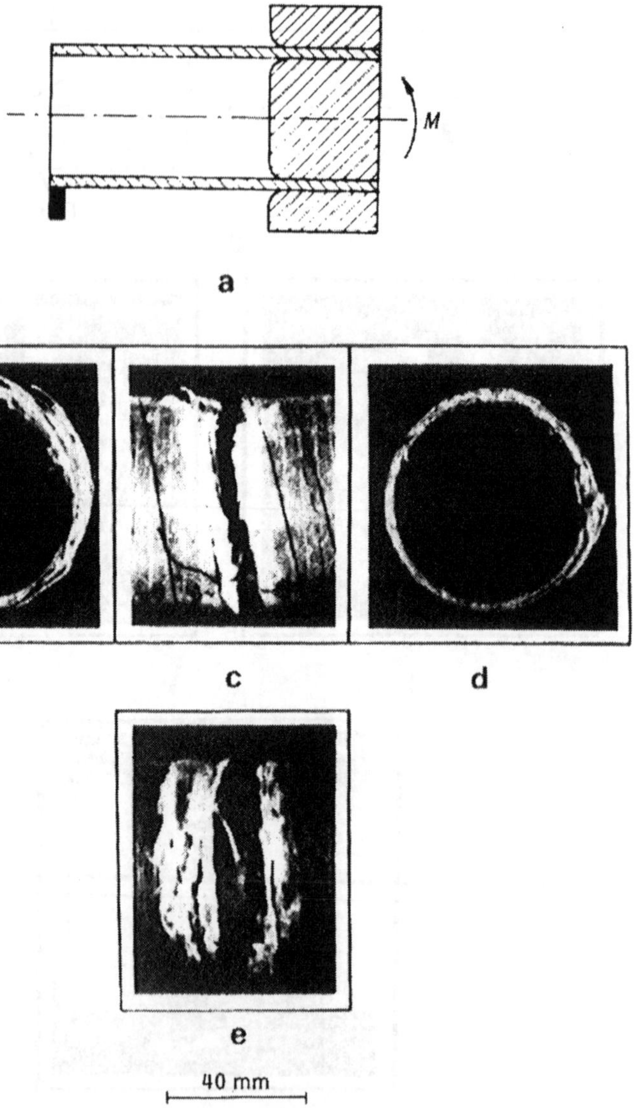

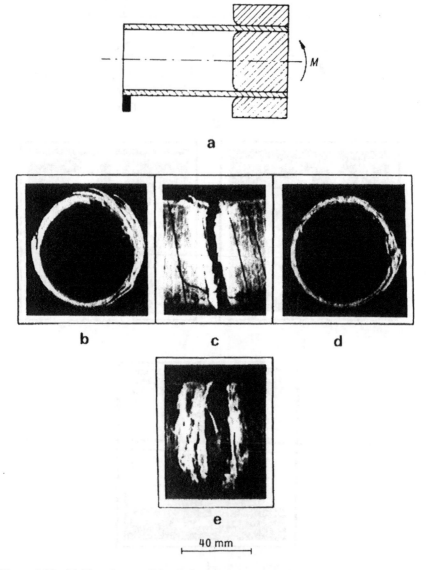

Figure 5.12. (a) Clamping condition D for specimen 6 (see Table 5.3), (b) left view of the cross section of the crushed region, (c) front view of the crushed region, (d) right view of the cross section of the crushed region, (e) compression zone.

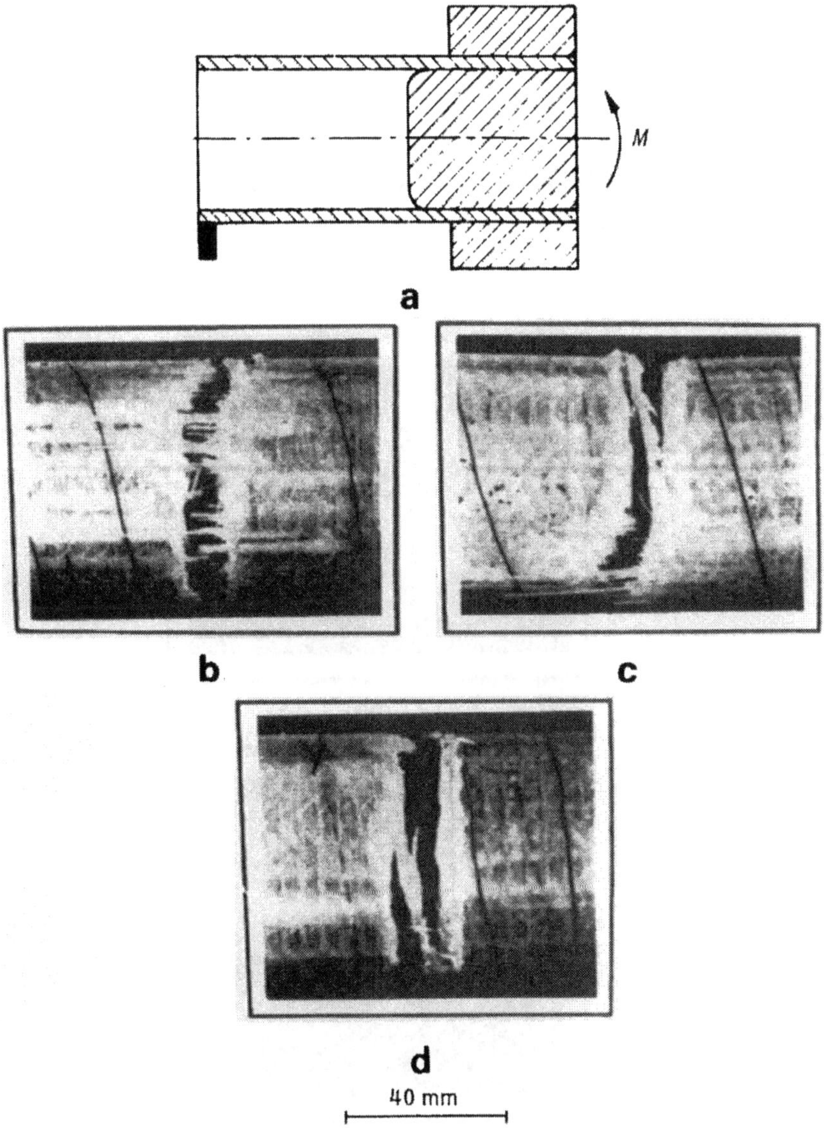

Figure 5.13. (a) Clamping condition E for specimen 7 (see Table 5.3), (b) tension zone, (c) front view of the crushed region, (d) compression zone.

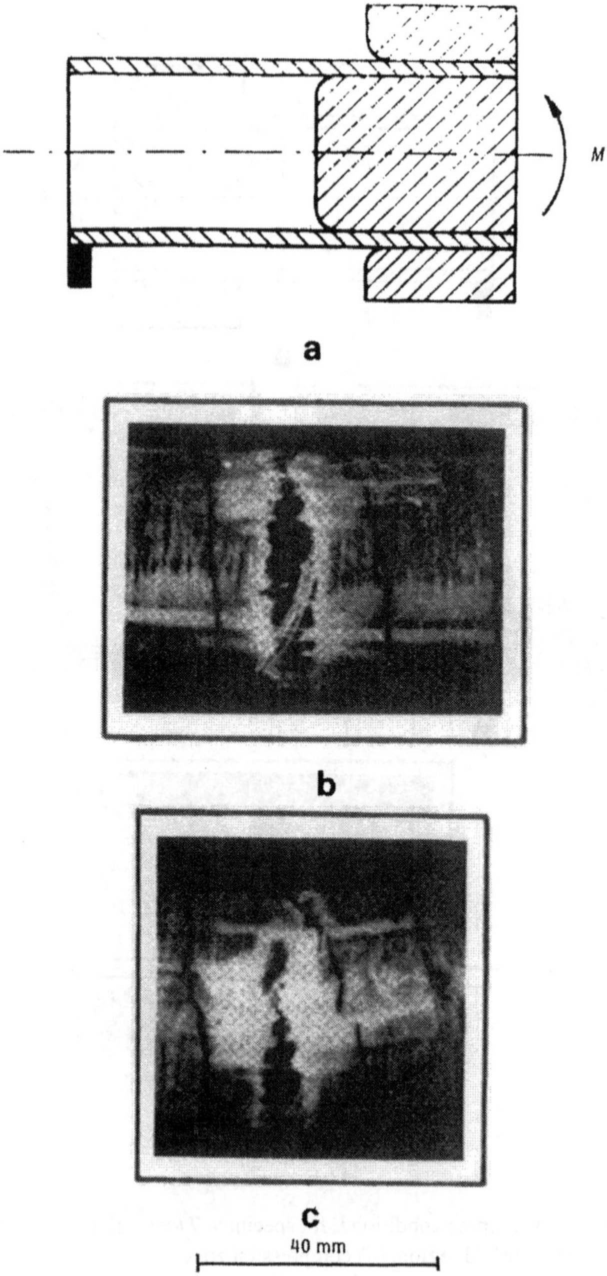

Figure 5.14. (a) Clamping condition F for specimen 8 (see Table 5.3), (b) tension zone, (c) front view of the crushed region.

Table 5.3 Crushing characteristics of circular tubes subjected to bending.

Specimen No.	Mean diameter $\bar{D}$ (mm)	Wall thickness t (mm)	Axial length L (mm)	Clamping condition	Peak bending moment M_{max} (N m)	Energy absorbed W (J)
1	56.0	2.3	256	A	773	286
2	56.0	2.3	256	C	954	160
3	56.0	2.3	256	A	702	160
4	56.0	2.3	256	C	921	135
5	56.0	2.3	256	B	912	232
6	56.0	2.3	256	D	1235	189
7	56.0	2.3	256	E	1497	191
8	38.0	2.3	256	F	482	65
9	38.0	2.3	256	A	391	133
10	38.0	2.3	256	C	500	77
11	38.0	2.3	256	B	355	64
12	38.0	2.3	256	D	364	52

400, 600 and 1200 grit abrasive wheels and polished using quarter micron alumina paste. In order to examine the microfracture developed, the finally prepared surfaces were investigated and photographed in a UNIMET metallographic optical microscope.

5.4.2 Failure Mechanisms: Experimental Observations

MACROSCOPIC OBSERVATIONS

Characteristic terminal views of the deformed specimens, showing the various collapse modes and macroscopic cracking details, are presented in Figures 5.9–5.14, whilst micrographs of the main microfractures, developed through the bending process, are shown in Figures 5.15–5.18.

The mode of collapse of bent tubes depends upon a number of factors, mainly related to the arrangement of the fibres, the properties of the matrix and fibres of the composite materials, the stacking sequence, the specimen size and the fixture device used at the clamped end of the tube.

Two distinct regions, with different macroscopic characteristics, are apparent through the bending process: the upper zone, mainly subjected to compressive loading and the lower one, under tensile straining, see Figures 5.9–5.14. Note that a narrow transition zone between them, with combined features, has been also observed. The fracture mechanisms governing the above two zones are quite different.

Collapse initiates in the compressive zone, close enough to the clamping device. In this region, the strength is mainly dominated by local fibre buckling stability, which depends on fibre diameter and modulus, and on the support given by the matrix. The associated forces result in a high stress field on the matrix material undergoing compressive failure by slipping and/or fracturing. In this way, a de-confinement of fibres

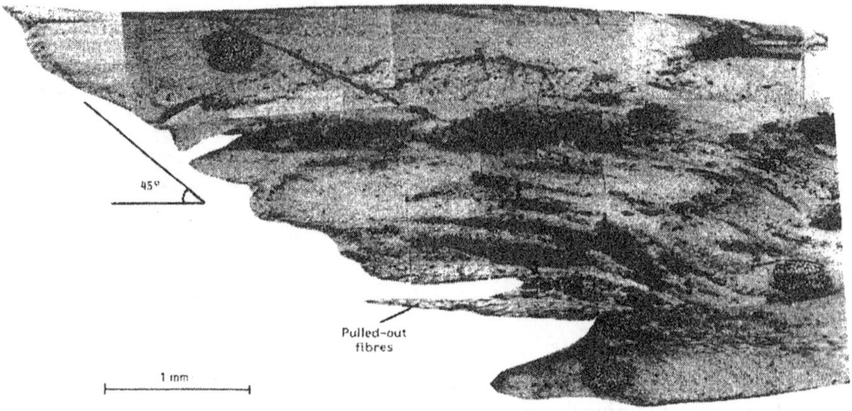

Figure 5.15. Micrograph showing microfailures of the compression zone of specimen 11 (see Table 5.3).

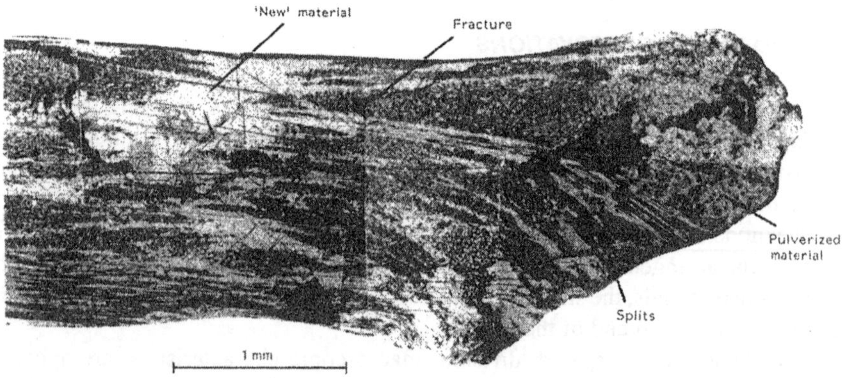

Figure 5.16. Micrograph showing microfailures of the compression zone of specimen 12 (see Table 5.3).

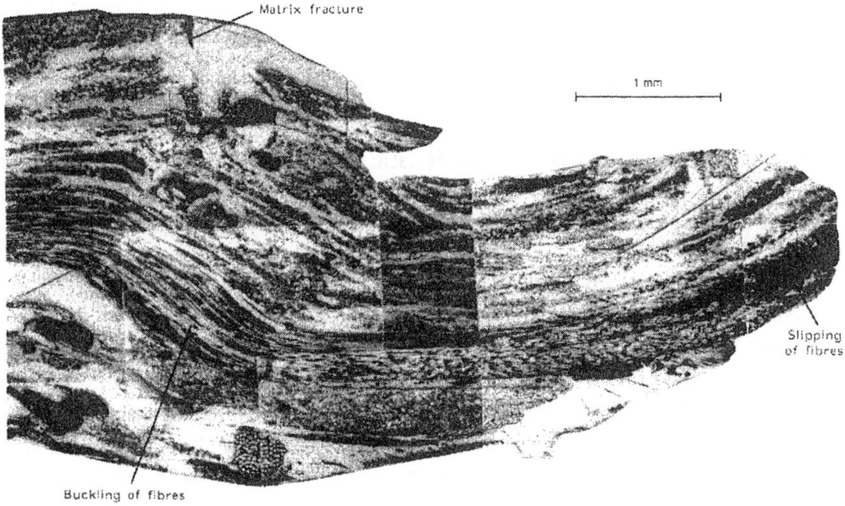

Figure 5.17. Micrograph showing microfailures of the compression zone of specimen 1 (see Table 5.3).

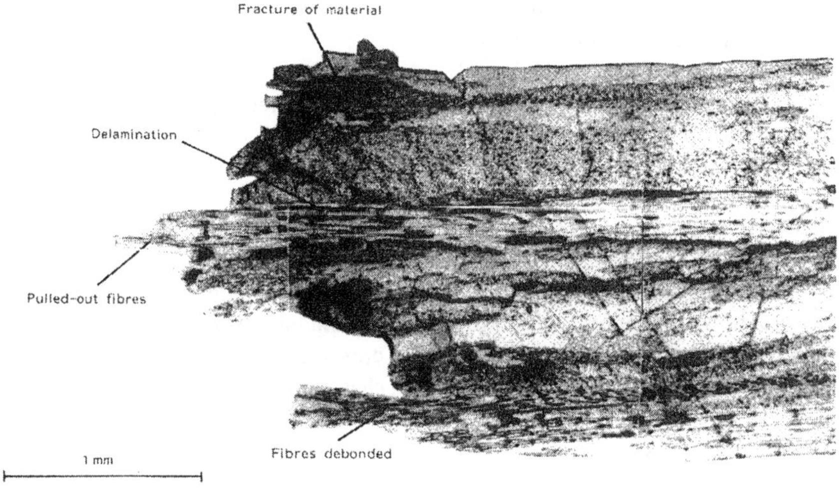

Figure 5.18. Micrograph showing microfailures of the tension zone of specimen 11 (see Table 5.3).

occurs leading to an increase in curvature and to a relaxation with an associated decrease in fibre forces and ultimate failure.

In the tensile region, damage initiates and propagates in zones of high stress concentrations, such as delamination edges etc., where micro- and macro-cracks first develop in the matrix phase. This matrix damage, however, does not imply to catastrophic failure. On the contrary, fracture occurs only when the fibre phase in the loading direction is sufficiently overstressed to reach fibre fracture strength. Because of the broken fibres and the load release, their neighbouring fibres are loaded by shear through the matrix over a characteristic stress transfer length, leading, therefore, to stress concentrations in the fibre phase; this tends to be relieved by matrix cracking parallel to the fibres, which in turn is stabilised by confinements introduced by the transverse plies.

The failure mechanism described above, which introduces significant strains perpedicular to the fibres, causes transverse cracking. The main features of this fracture are greatly affected by the fixture conditions at the clamped end of the tube.

A plug inserted into the inside diameter of the tube usually leads to full separation of the shell in two parts, with main characteristics the pulled-out fibres giving a "brush-like" appearance of the tensile region, see Figures 5.11(d) and 5.13(b), and a considerably smoother surface of the compression zone with short crushed fibre ends and much debris, see Figures 5.9–5.14.

In the cases of simple clamping (clamping conditions A and B), see Figures 5.9 and 5.10, during the bending process, the upper shell region bends inwards showing deformation characteristics similar to those observed for steel tubes, see Figures 5.9(d) and 5.10(d) and the remarks reported in Reference [102]. Transverse cracking, occurred at the upper region, usually propagates towards the middle of the tube circumference without causing tube separation.

In many cases, a longitudinal crack, accompanied by fibre-matrix debonding and severe delamination, propagates in the longitudinal direction at the outer region of the compression zone, probably due to the folding of the separated parts of the tube towards to each other, see Figures 5.10 (b) and 5.12 (b).

MICROSCOPIC OBSERVATIONS

Micrographs of sections of deformed regions in the longitudinal direction of the tube are shown in Figures 5.15–5.18. Due to the random orientation of fibres, they are characterised as, "transverse", with circular cross-section mainly in the hoop direction, or "longitudinal" fibres, with elliptical shape in the 0° orientation (i.e. the tube axis).

Regions under compressive loading are mainly characterised by, see Figures 5.15–5.18:

- Cracking, which in most specimens is developed at an angle of about 45° with reference to the tube axis; see Figure 5.15.
- Fracture across the fibres, resulting in considerably smoother surface than that under tensile loading; see below and compare Figures 5.15–5.17 with Figure 5.18. This is

mainly caused by micro-buckling of fibres in the form of a characteristic shear band across the fibres. Transverse cracking usually propagates through the boundaries between fibres of different orientation; see Figures 5.15–5.17.

- Regions with small fibre volume fraction, which are more sensitive to fracture, causing initial local fold during the bending process; see Figures 5.15 and 5.17.
- Parts with initially incomplete resin layer followed by subsequent resin completion; fracture initiates in the boundaries between "old" and "new" matrix material, see Figure 5.16.
- Characteristic cracking of fibres and resin, as well as bending and/or buckling of longitudinal fibres without fracture; see Figures 5.15–5.17.
- Slipping of fibre groups of different orientation, with simultaneous microfragmentation of the in-between region; see Figure 5.17.
- Splitting parallel to fibres, due to corresponding low energy fracture path, see Figure 5.16. Fracture across the fibres requires a considerably greater amount of energy than splitting parallel to the fibres. Although the main fracture path is in a direction across the fibres, the fracture consists of many minor splits longitudinally directed for a certain length over which the fibres are debonded from the resin.

Tube regions, loaded in tension, exhibit complex non-homogeneous damage and failure modes, such as, see Figure 5.18:

- Matrix cracking
- Fibre/matrix interface debonding, mainly for transverse fibres
- Delamination and fibre breakage, mainly of longitudinal fibres

The severity of damage depends on the laminar strength and stiffness, the stacking sequence, the ply orientation, the fibre volume fraction and the specimen size.

5.4.3 Energy Absorbing Characteristics

Bending moment, M vs. angle of rotation, θ curves were obtained automatically, as described above, and typical M/θ graphs are presented in Figures 5.19–5.20, corresponding to the various clamping conditions. The maximum bending moment, M_{max} and the absorbed energy (i.e. the area under the M/θ curve) are measured and tabulated in Table 5.3. The following remarks, concerning the strength and energy absorbing capacity of the tube during the bending process, may be drawn.

During the bending process, the tube initially deforms elastically and the M/θ curve is characterised by a sharp steady-state increase of the bending moment until a maximum value, M_{max} is attained. Cracking occurs close to clamping end; the plastic post-crushing regime follows, where the initially developed transverse crack spreads gradually or rapidly over the whole cross-section of the tube, mainly depending upon the clamping conditions, see Figures 5.19–5.20.

The insertion of a plug, as additional clamping device, leads to higher initial maximum values of the bending moment, with simultaneous shortening of the post-

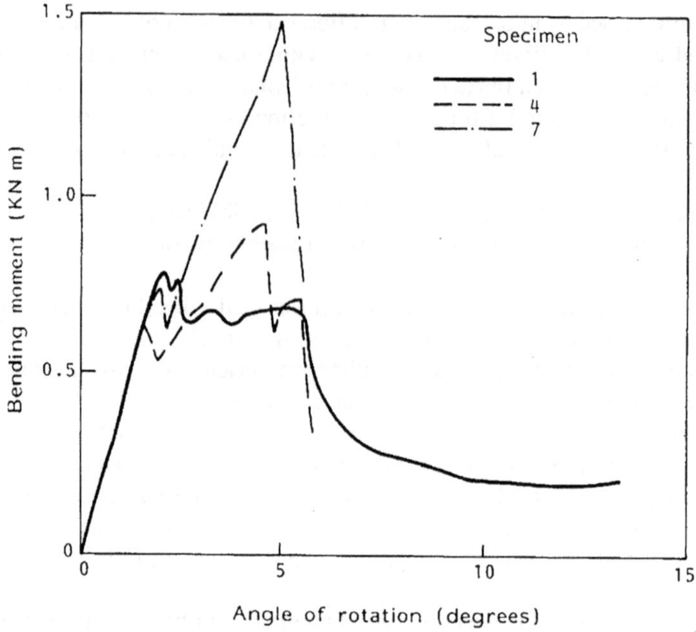

Figure 5.19. Bending moment/angle of rotation characteristics of bent tubes (see Table 5.3 for details).

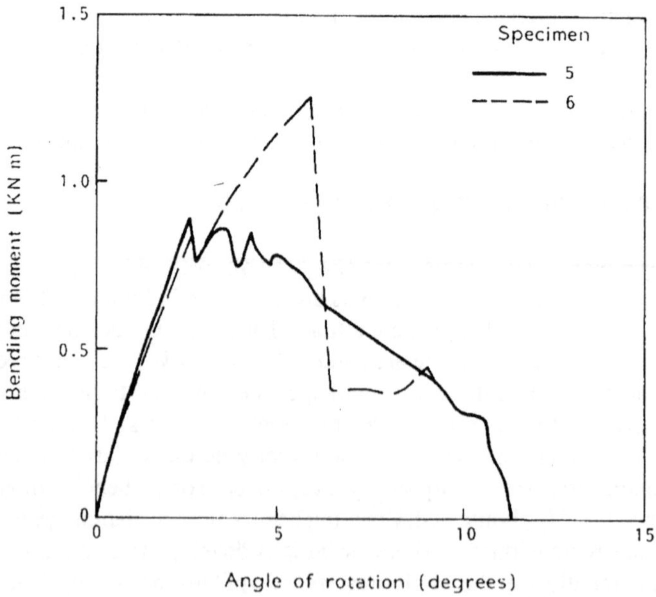

Figure 5.20. Bending moment/angle of rotation characteristics of bent tubes (see Table 5.3 for details).

crushing regime, probably due to the acceleration of the crack propagation, resulting in lower amounts of energy dissipated; compare specimens 1, 4 and 7 in Figure 5.19 and see also Table 5.3. Note that the length of the plug plays also a role in the collapse efficiency (rigidity) of the bent tube, resulting in a significant increase of both the initial maximum bending moment and the energy absorbing capacity; compare specimens 4 and 7 in Figure 5.19 and see also Table 5.3.

Clamping devices with rounded edges delay the crack development and propagation, causing, in general, higher values of bending moment and amounts of energy absorbed; see Figures 5.19 and 5.20 for specimens 1 and 5 or for specimens 4 and 6.

5.4.4 Crashworthy Capability: Concluding Remarks

The crush behaviour of thin-walled circular tubes made of composite material and subjected to bending at various end-loading conditions was experimentallly investigated, leading to the classification of the main fracture and energy absorbing characteristics of the tested specimens during the bending process.

As clearly indicated, the composite material tubes are deformed in a manner quite different than tubes of similar geometry made from conventional metallic materials, [102]; the former do not undergo plastic deformation but collapse is mainly due to extensive microcracking which depends upon the properties of fibres and resins as well as upon the fibres orientation.

Two distinct zones of the tube cross-section close to the clamping device, under compression and tensile straining conditions respectively, were apparent. The fracture characteristics of each region are different and depend upon the loading and clamping conditions, and they are greatly influenced by the existence and the size of the inserted plug.

The collapse mechanism of the composite material tube is complex and strongly influenced by the type of fibres and matrix system employed and the nature and strength of the fibre-matrix interface bond. Fracture of matrix and fibres, debonding of the fibres from the matrix and friction, required to pull broken fibres from the matrix, must be taken into account. The energy absorbing efficiency is also affected by the clamping conditions. Inserting a plug, the post-crushing region is shortened, resulting in lower amounts of energy dissipated; an increase of energy dissipated when round-ended clamping rings are used is obtained.

The strength characteristics, expressed by the peak bending moment, are influenced by the presence of the inserted plug, leading, in general, to higher initial peak values of bending moment, whilst rings with rounded edges delay the crack development and propagation, causing in this manner even higher values of bending moment.

CHAPTER 6

SQUARE/RECTANGULAR TUBES

6.1 NOTATION

b = external side width of square tube
d_1 = height of rectangular tube cross section
d_2 = width of rectangular tube cross section
G = fracture toughness
k = constant
L = axial length of square tube
L_c = length of central crack
l_s = side length of pulverised wedge
M = bending moment
M_{max} = maximum (peak) bending moment
n = number of axial splits
P = current crushing load
$\overline{P}$ = mean crushing load
P_1, P_2 = normal force per unit length
P_{max} = peak load
R_{ad} = fracture energy per unit area of layers
s = displacement, shell shortening, crush length
t = wall thickness of square tube
v = crush-speed
W = energy absorbed
W_S = specific energy
W_T = total energy dissipated
W_{tr} = energy required for the crush zone formation
α = angle of wedge
ε = strain
θ = angle of rotation
θ_{max} = angle of rotation corresponding to M_{max}

θ^* = angle between axis 1 and x
μ_s = static friction coefficient
μ_d = dynamic friction coefficient
σ_θ = tensile fracture stress
σ_o = normal stress
$\varphi\ (=\alpha/2)$ = semi-angle of wedge

6.2 GENERAL

In the present chapter are reported the behaviour and crashworthiness characteristics of square composite tubes, subjected to static and dynamic axial loading exerted by a hydraulic press and a drop-hammer, respectively [76]. The effect of specimen geometry, i.e thickness and axial length, and of the loading rate on the energy absorbing capability are studied in detail. Attention is directed towards the mechanics of the axial crumbling process, from macroscopic and microscopic point of view, for facilitating engineering design calculations of the amount of energy dissipated. A theoretical analysis of the collapse mechanism of the components tested under axial compression is proposed, leading to a good approximation of the energy absorbed during crushing.

Furthermore, the crashworthy behaviour of cantilever thin-walled composite tubes of square and rectangular cross-section during bending is studied experimentally, both macro- and microscopically [33,96]. Investigations concerning the influence of the specimens geometry and the process conditions are reported; comparison with equivalent circular tubes from the absorption efficiency point of view is also made and discussed. Moreover, a theoretical analysis for the prediction of the ultimate bending strength is presented.

6.3 AXIAL COLLAPSE: STATIC AND DYNAMIC

6.3.1 Experimental

Two different kinds of fibreglass composite material, designated as composite materials A and B, respectively, were used for testing. Material A was a fibreglass composite material with individual fibre diameter of 9 μm chopped strand mat with random fibre orientation in the plane of the mat. The shells were fabricated by a hand lay-up technique using pieces of fibreglass cloth (0.8 g/mm^2) and impregnating it with a polyester resin, providing in this manner with a composite material of 72% per weight fibre content and 1.37 g/cm^3 density. Material B was a commercial glass fibre and vinylester composite. The tube wall consisted of nine plies with a total thickness of 2.5 mm for the axial collapse specimens. Starting from the exterior of the shell the plies were laid-up in the sequence $[(90/0/2R_c)/(2R_c/0/90)/R_{c.75}]$, where the 0° direction coincides with the axis of the tube; R_c denotes random chopped strand mat plies and $R_{c.75}$ represents a similar ply but thinner, providing in this manner with a composite material of 33.9% per

per volume fibre content and 1.55 g/cm³ density. According to the manufacturer's specifications, the laminate fibre lay-up was hand wrapped around a rigid foam core, whilst plastic staples were used to hold the lay-up on the core during the resin transfer molding process. Details about the fabrication of the composite tube are presented in Chapter 5. Stress-strain curves, as obtained from quasi-static tension testing, for both materials A and B, are shown in Figure 5.4 in Chapter 5.

The static axial collapse was carried out between the parallel steel platens of a SMG hydraulic press at a crosshead spead of 10 mm/min or a compression strain-rate of 10^{-3} sec^{-1}. The corresponding dynamic tests were performed by direct impact on a drop-hammer at velocities exceeding 1 m/s. The existing drop-hammer facility, with a 35 kg falling mass from a maximum drop height of 4m, provides a maximum impact velocity of about 10 m/s. The experimental set-up and measuring devices used throughout these tests are described in detail in Chapter 5. Load/tube shortening (displacement) curves during the crushing process were automatically measured and recorded for both types of loading. The values of the initial peak load, P_{max} and the energy absorbed, W for the axially collapsed specimens, obtained by measuring the area under the load/displacement curve, as well as the mean crushing load, $\overline{P}$ (defined as the ratio of energy absorbed to the total shell shortening) and the specific energy, W_s (equal to the energy absorbed per unit mass crushed, calculated as the crushed volume times the density of the material) are tabulated in Table 6.1.

Photographs showing characteristic terminal views of the deformed specimens for all series of experiments are presented in Figure 6.1. Typical micrographs of the crush zone showing the main microfailures, as obtained using a Unimet metallographic optical microscope, are also shown in Figure 6.2. For details about specimens preparation and polishing for the microscopic observations see Chapter 5.

6.3.2 Failure Mechanisms: Experimental Observations

When load is applied to the edges of the tube, local failure of material occurs and small inter/intralaminar cracks are formed. The length of the inter/intralaminar cracks and, whether the lamina bundles fracture, determines whether the resulting crushing mode is transverse shearing, lamina bending or a combination of these modes (brittle fracturing) [66].

Two distinct modes of collapse, classified as Mode I and Mode III, respectively, were observed throughout the axial static and dynamic tests. Short specimens up to a certain length, followed the progressive collapse mode of failure (Mode I), whilst relatively long specimens are characterised by the failure Mode III, i.e. the column-buckling type of collapse; see the relevant remarks in Reference [76].

PROGRESSIVE END-CRUSHING (MODE I)

The Mode I of failure, similar to a "mushrooming" failure, is characterised by progressive collapse through the formation of continuous fronds which spread outwards and inwards, see Figures 6.1(a) and (b) and References [64, 98, 93, 103]. As deformation proceeds further, the externally formed fronds curl downwards, with the simul-

Table 6.1: Crushing characteristics of axially loaded square tubes

(a) Static

Sp. No	Mater.	Num. of layers	Thick-ness, t (mm)	Axial length, L (mm)	Side width, b (mm)	Collapse mode	Crush length, s (mm)	Crushing load, P (kN) Initial peak, P_{max} Exper.	Mean post-crushing, $\bar{P}$ Exper.	Mean post-crushing, $\bar{P}$ Theor.	Total energy absorbed, W_T (kJ) Exper.	Total energy absorbed, W_T (kJ) Theor.	Specific energy, Ws (kJ/kg) Exper.	Specific energy, Ws (kJ/kg) Theor.
1	A	2	2.5	103.8	48.0	Ia	37.7	22.3	25.3	27.2	0.954	1.025	40.6	41.2
2	A	3	3.3	101.8	47.6	Ia	38.7	61.7	32.6	34.9	1.262	1.350	40.7	41.6
3	B	*	2.5	50.8	47.7	Ia	32.5	65.2	39.5	41.3	1.284	1.342	50.1	50.3
4	B	*	2.5	76.2	47.7	Ia	48.5	63.1	40.1	41.3	1.945	2.003	46.7	50.3
5	B	*	2.5	101.6	47.7	Ia	63.5	64.8	40.6	41.3	2.578	2.623	47.3	50.3
6	B	*	2.5	152.4	47.7	Ia	63.5	64.4	39.7	41.3	2.521	2.623	46.3	50.3
7	B	*	2.5	203.2	47.7	Ia	63.5	66.1	41.2	41.3	2.616	2.623	50.3	50.3
8	B	*	2.5	304.8	47.7	Ia	63.5	53.4	41.6	41.3	2.642	2.623	50.7	50.3
9	B	*	2.5	304.8	47.7	III	25.4	65.3	-	-	0.338	-	-	-
10	B	*	2.5	444.5	47.7	III	25.4	67.6	-	-	0.249	-	-	-

Table 6.1 (cont.)

(b) Dynamic

Sp. No	Mater.	Num. of layers	Thick-ness, t (mm)	Axial length, L (mm)	Side width, b (mm)	Crush speed, v (m/s)	Collapse mode	Crush length, s (mm)	Crushing load, P (kN) Initial peak, P_{max} Exper.	Mean post-crushing, $\bar{P}$ Exper.	Mean post-crushing, $\bar{P}$ Theor.	Total energy absorbed, W_T (kJ) Exper.	Total energy absorbed, W_T (kJ) Theor.	Specific energy, Ws (kJ/kg) Exper.	Specific energy, Ws (kJ/kg) Theor.
11	A	2	2.3	106.5	45.4	6.0	Ia	49.6	30.7	17.0	19.2	0.844	0.952	27.0	28.6
12	A	3	3.1	105.5	45.6	6.0	Ia	30.0	82.1	24.7	26.6	0.740	0.798	34.2	35.5
13	A	2	2.5	105.9	46.4	7.0	Ia	60.3	20.2	17.7	19.5	1.068	1.176	29.5	31.0
14	A	3	3.3	107.2	49.0	7.0	Ia	33.0	70.6	30.9	32.2	1.020	1.063	33.2	34.0
15	A	3	3.5	102.7	48.4	8.1	Ia	45.7	91.9	31.7	32.8	1.450	1.499	35.0	36.2
16	B	*	2.5	50.8	47.7	7.0	Ia	32.5	70.1	51.5	48.8	1.674	1.586	62.8	60.2
17	B	*	2.5	50.8	47.7	8.1	Ia	42.6	72.6	50.6	48.8	2.156	2.079	61.8	60.2
18	B	*	2.5	76.2	47.7	7.0	Ia	32.7	72.7	49.7	48.8	1.626	1.596	60.7	60.2
19	B	*	2.5	76.2	47.7	8.1	Ia	42.8	74.5	51.8	48.8	2.217	2.089	63.2	60.2
20	B	*	2.5	101.6	47.7	7.0	Ia	32.4	69.3	50.5	48.8	1.636	1.581	61.6	60.2
21	B	*	2.5	101.6	47.7	8.1	Ia	42.1	71.5	51.1	48.8	2.151	2.055	62.4	60.2
22	B	*	2.5	152.4	47.7	7.0	Ia	33.1	73.2	50.9	48.8	1.685	1.615	62.1	60.2
23	B	*	2.5	152.4	47.7	8.1	III	62.5	77.6	-	-	0.310	-	-	-
20	B	*	2.5	203.2	47.7	7.0	III	55.6	76.7	-	-	0.292	-	-	-

* See Section 5.3.1 for material B lay-up

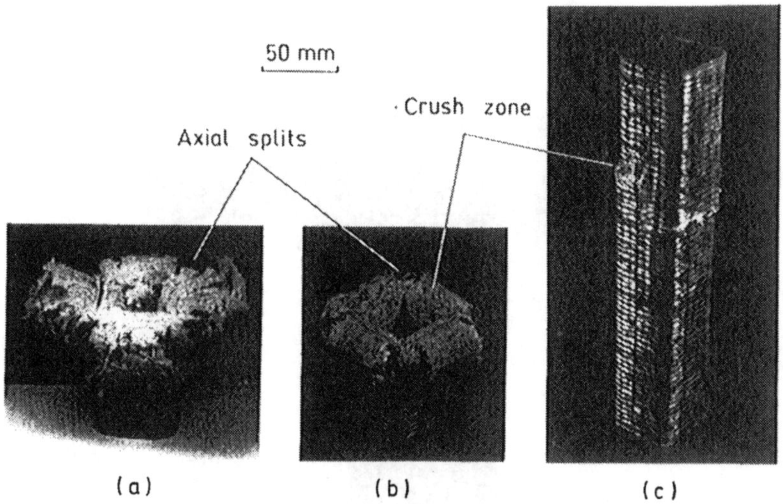

Figure 6.1. Terminal microscopic views of (a) Mode I of collapse (sp. 2; material A). (b) Mode I of collapse (sp. 5, material B). (c) Mode II of collapse (sp. 9) (see Table 6.1).

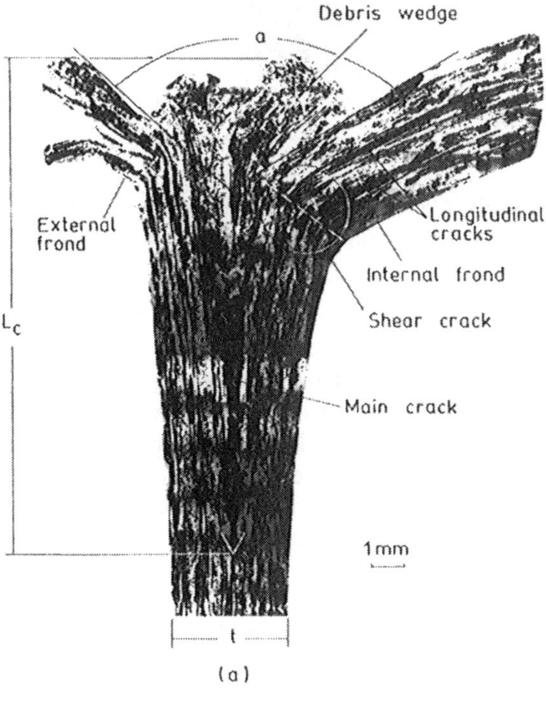

Figure 6.2. Micrographs showing microfailures in the crush zone in the middle of the tube side for square tubes: (a) statically loaded (sp. 2; material A).

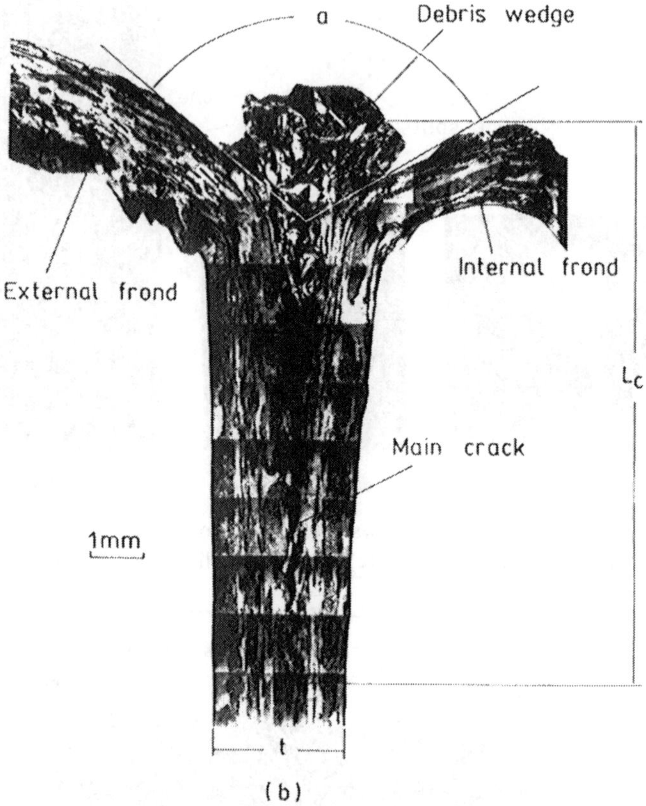

Figure 6.2 (continued). Micrographs showing microfailures in the crush zone in the middle of the tube side for square tubes: (b) statically loaded (sp. 5; material B).

taneous development of four axial splits followed by splaying of the material strips. Note that the splits are always located at the four corners of the shell, probably due to local stress concentration during the very early stage of straining. Axial tears were not apparent in the internal fronds, which were more continuous than their external counterparts. At the early stage of loading, the shell initially behaved elastically and, as soon as the load attained a peak value, depending on shell geometry, material characteristics and corners rigidity, cracks formed at each of the four corners and propagated downwards along the tube axis, see Figures 6.1(a) and (b); they are associated with the formation of a main central intrawall crack at the end of the shell adjacent to the loading area, see Figure 6.2.

The post-crushing regime is characterised by the formation of two equal lamina bundles bent inwards and outwards due to the flexural damage, which occurs at a distance from the contact surface equal to the wall thickness; they withstand the applied load and buckle when the load, or the length of the lamina bundle, reaches a critical value. At this stage, a triangular debris wedge of pulverised material starts to form,

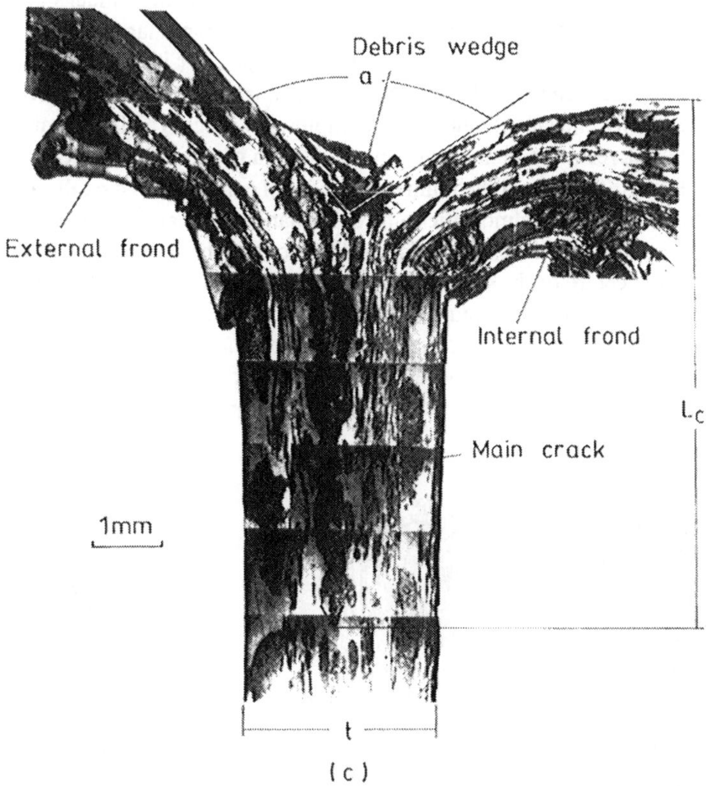

Figure 6.2 (continued). Micrographs showing microfailures in the crush zone in the middle of the tube side for square tubes: (c) dynamically loaded (sp. 17; material B) (see Table 6.1).

see Figure 6.2; its formation may be attributed to the friction between the bent bundles and the platen of the press or the drop-mass. As loading proceeds further, resulting in crushing with the subsequent formation of the internal and external fronds, normal stresses develop on the sides of the debris wedge followed by shear stresses along the same sides due to the friction at the interface between the wedge and the fronds. Note, also, that additional normal and shear stresses develop at the interface between the steel press platen or the drop mass and the deforming shell as the formed fronds slide along this interface.

As observed [63, 75], the most common case for a circular tube under stable collapse mode is characterised by the formation of a central interlaminar opening crack at the apex of a highly pulverised wedge of constant shape over its circumference, which separates the tube wall in two lamina-bundles of equal thickness. The debris wedge remains essentially unchanged during the compression process and penetrates the composite material, resulting in the development of high frictional resistance of it with the adjacent fronds. The size of the main crack is small compared to the tube axial length. Square tubes developed almost the same deformation mechanism to that observed for circular tubes, but with the characteristic difference that the central crack length diminishes from the centre of the square side towards the square corners, where the typical crush zone disappears, see Figure 6.3. The maximum value of the crack length, L_c, which is attained at the middle of each side of the square cross section, is almost the same as the corresponding one observed in the case of circular tubes loaded under the same conditions. Based on the above mentioned micromechanism, as well as on secon-

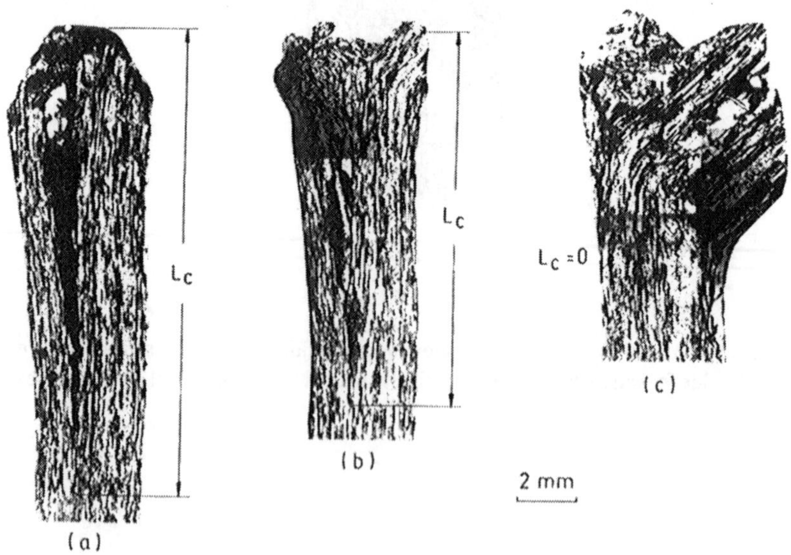

Figure 6.3. Micrographs showing microfailures in the crush zone for statically loaded tube (sp. 2; see Table 6.1) in positions located at (a) the middle of the side of the square cross section, (b) the 1/4 of the side length, (c) the corner of square cross section.

dary failure mechanisms contributing to the overall energy absorption during collapse, the following principal sources of energy dissipation at microscopic scale may be listed: Intrawall crack propagation; fronds bending due to delamination between plies; axial splitting between fronds; flexural damage of individual plies due to small radius of curvature at the delamination limits; frictional resistance to axial sliding between adjacent laminates; frictional resistance to the penetration of the debris wedge; frictional resistance to fronds sliding across the platen.

MID-LENGTH COLLAPSE MODE (MODE III)

Specimens followed this collapse mode exhibited extensive brittle fracture along their circumference, see Figure 6.1 (c). Fracture started at a distance from the loaded end of the specimens, approximately equal to the mid-height of the shell, and involved catastrophic failure by cracking and separation of the shell into irregular shapes, probably due to local severe shear straining of the wall of the shell. This failure mode is similar to the Euler column-buckling of very thin metallic and PVC tubes subjected to axial loading [8].

As outlined above, two distinct modes of collapse, Mode I and Mode III, were observed throughout the axial static and dynamic tests, concerning square tubes. Note that only the Mode I of collapse was obtained for the tubes made of material A due to their short axial length. Short specimens, up to a certain length, followed the Mode I of collapse, whilst relatively long specimens are characterised by the Mode III. Note that for impact loading the Mode III occurred for shorter shells; compare the crushing characteristics for static and dynamic loading presented in Table 6.1. As far as the Mode I of collapse is concerned, the initial axial length of the shell, L seems not to affect the length of the main central crack, L_c and the mechanical response of the shell due to failure, see Table 6.1. Dynamic loading greatly affects the microfracture mechanism outlined above. The size of the debris wedge and the main crack dimensions, caused by the wedge propagation, are smaller in impacted shells, as compared to statically loaded ones; compare Figures 6.2(b) and (c). Note that all the above mentioned characteristics, concerning the behaviour of the square tubes made of material A and B under axial loading, are in agreement with those reported in Chapter 5 for circular tubes made of the same material and loaded under the same conditions.

6.3.3 Energy Absorbing Characteristics

Typical load/displacement curves for each mode of collapse for static and dynamic loading of both geometries are shown in Figure 6.4. Initially the shell behaves elastically and the load rises at a steady rate to a peak value, P_{max} and then drops abruptly. As deformation progresses, the shape of the load/displacement curve is associated entirely with the mode of collapse. For thin-walled composite shells subjected to axial loading, it was observed that the fracture behaviour of the shell appears to affect the loading stability, as well as the magnitude of the crush load and the energy absorption during the crushing process. It may be assumed that, at any instant, the crush load must be supported by some, more than one, structural elements of the shell and,

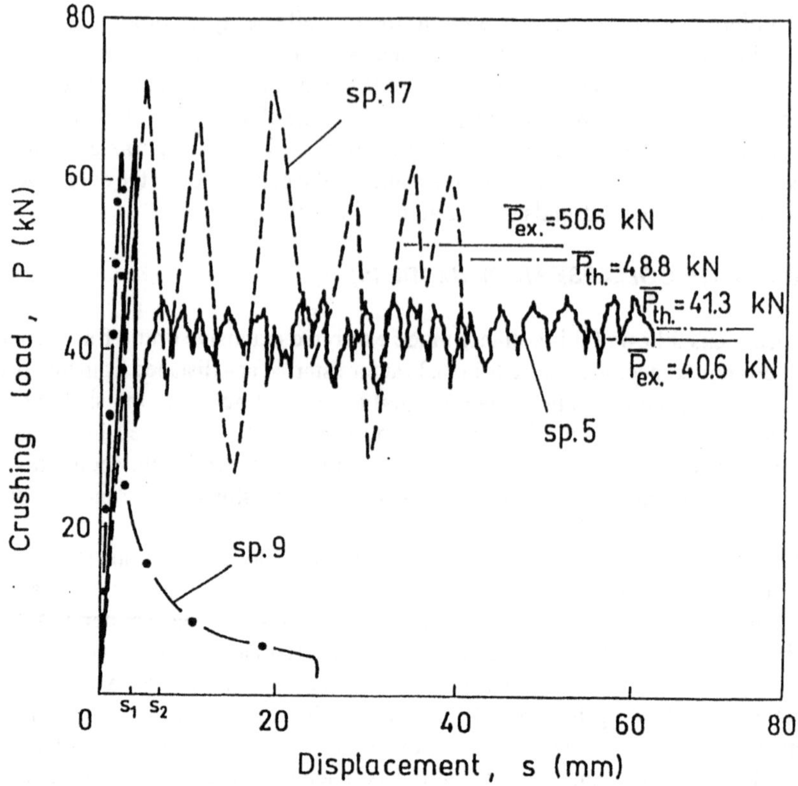

Figure 6.4. Load/displacement curves for square tubes subjected to static and dynamic loading for various collapse modes (—Mode I, static (sp. 5); —•—Mode II, static (sp. 9); --- Mode I, dynamic (sp. 17))(see Table 6.1).

furthermore, a specific element contributes more to supporting the load, see also Reference [84]. At the stage that crush load reaches the peak value, P_{max}, and for the Mode I of collapse, cracks form at each of the four corners accompanied by the formation of a circumferential intrawall crack at the end of the shell adjacent to the loading area; the post-crushing region is characterised by oscillations about a mean post-crushing load, $\overline{P}$. For the case of the progressive end-crushing Mode I, the load/displacement curves showed several similar features for both materials.

Dynamically obtained load/displacement curves show more severe fluctuations, with troughs and peaks, than the corresponding statically ones, see also Figure 6.4. From the shape of the dynamically obtained curves at the post-crushing region, it rendered difficulties in assuming the possible fracture mechanism occurred, as well as the development and propagation of the microcracks. On the contrary, for a static test, the deformation mode encountered during collapse could be excluded from the pattern of a load/displacement curve (history of deformation). Moreover, due to the dynamic nature of the phenomenon (duration ranges between 7–25 ms), the sequence of different microcracking processes in relation to the shape of the obtained load/displacement curve can not be distinctly followed.

The load/displacement curves, corresponding to the mid-length collapse mode (Mode III), showed a typical pre-crushing region, but the initial elastic response was followed by a very sharp drop in load and poor post-crushing characteristics, see Figure 6.4. The Mode III is similar to that of the Euler column-buckling mode obtained when axially compressing very thin metal or PVC tubes, offering very small energy absorbing characteristics [8].

It was shown that for end in-plane loaded flat plates, most of the load is carried at the edges of the plate, whilst in the case of loaded square tubes at their corners [104]. Likewise, it would be expected that the present square tubes would support most of the load at their corners. Therefore, the support of the surface mats depends on the integrity of the corners and, subsequently, they would fail before the strength of the material in the straight section would reach a critical crushing stress. As crushing progresses, and if the dimensions of the material are within a certain range, a bending moment at some distance away from the top edge of the side would reach a critical value and the material would fail by bending with a number of irregular undamaged segments; see the details of the Mode III of collapse in Figure 6.1(c).

As far as the initial maximum value of the load in the elastic region, P_{max} is concerned, it must be noted that this magnitude is greatly affected by the wall thickness and the mean circumference of the shells for both static and dynamic loading, see Table 6.1 for details; it is evident that the axial length of the tubes has no effect on the peak load. Note that, as far as the dynamically loaded shells of both materials is concerned, the peak load, P_{max} overestimates the corresponding static ones for the materials and the loading conditions examined. For a given material, the size of the elastic deformation region is greater for tubes displaying greater thickness and side width. This depends on the squareness and the trigger mechanisms of the square tube; e.g. if the top end is slightly beveled, the magnitude of load in the elastic region will be reduced due to the smaller amount of material being loaded at the press platen. The load reached at the end of the elastic region, for an untriggered shell undergoing the Mode I of collapse, is usually the maximum load that the shell may sustain. It must be noted that the peak load has very little effect on the mean load and the total energy absorbed.

From Table 6.1, it is evident that the mean post-crushing load and the energy absorbed are mainly affected by the crushing length, whilst the axial length of the shell has no significant effect on these crashworthy characteristics. For square tubes made of material B, higher values of the mean post-crushing load and the energy absorbed were predicted for dynamic collapse, see Figure 6.4 and Table 6.1, probably due to the higher values of the dynamic friction coefficients, μ_d, between the wedge and the fronds and between the dropped mass surface and the fronds (about 15–20% of the related static ones); this resulted in an increase of the crashworthy ability of the shell of about 20%. Note that similar observations were reported in Chapter 5, pertaining to the axial collapse of circular tubes, made of the same material as the square tubes and tested under the same conditions. From the results reported, it is indicated that the energy absorbing capability of the former is better than that of the latter ones, mainly due to the shell geometry and the fracture mechanism associated with it.

For the dynamically loaded tubes made of material A, the effect of the strain-rate on the specific energy and the mean post-crushing load seems to be almost negligible, see Table 6.1(b). For all specimens tested, the peaks and valleys are not identical, probably due to the stochastic sequence of the phenomena involved. From Table 6.1, it is evident that static tests for the same shell geometries developed higher values of specific energy than these obtained in dynamic testing, probably due to the higher values of the static friction coefficients, μ_s, between the wedge and the fronds and between the platen surface and the fronds (about 10–15% more than the related dynamic ones); this increase in the crashworthy ability of the shell was about 15%. This is in agreement with the results reported in Chapter 5, concerning the crashworthy characteristics of circular tubes made of material A, whilst it contradicts to the above mentioned remarks, pertaining to the square tubes made of material B and subjected to static and dynamic loading. Therefore, it may be concluded that the most important parameter, that determines the governing phenomena which affect the magnitude of the friction coefficients for the various testing conditions, is the shell material.

6.3.4 Failure Analysis

STATIC AXIAL COLLAPSE

Consider the failure mechanism, Mode I, outlined above. Various crashworthy phenomena, pertaining to the axial collapse of composite multi-layered shells, are associated with the distribution of the absorbed energy during the crushing process. In the proposed theoretical approach, see Figure 6.5, the following crushing phenomena are encountered: friction between the annular wedge and the fronds and between the fronds and the platen of the press; fronds bending; crack propagation; axial split-

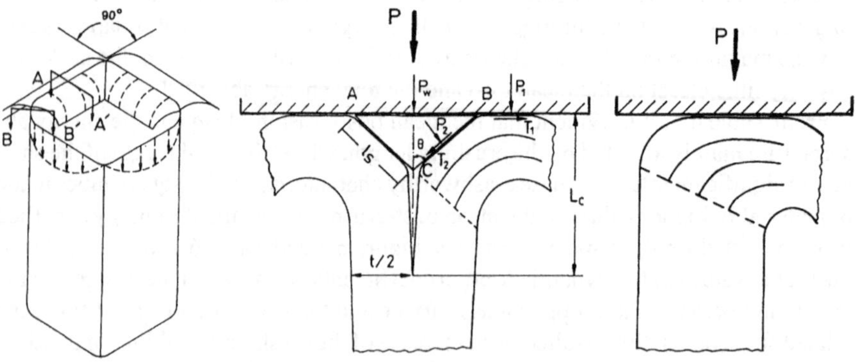

Figure 6.5. (a) Configuration of failure mechanism of Mode I, (b) configuration of the crush zone in the middle of the tube side (cross section AA′ of (a)), (c) configuration of the crush zone in the corner of the tube (cross section BB′ of (a)).

ting. The theoretical model proposed in Chapter 5, for the analysis of composite circular tubes subjected to static axial loading, was modified and used to analyse the collapse mechanism and to estimate the related energy absorbed during the axial crushing of the square tubes.

During the elastic deformation of the shell the load rises at a steady rate to a peak value, P_{max}, see Figure 6.4. At this stage, cracks of length L_c form at the four corners of the tube and propagate downwards along the tube axis, splitting the shell wall, see Figure 6.5(a). They are accompanied by the development of a circumferential central intrawall crack at the top end of the shell; the related shell shortening is s_1, see Figure 6.4. It is assumed that the crack length distribution along the circumference of the tube follows an elliptical configuration as shown in Figure 6.5(a). The maximum value of the crack length, L_c is attained at the middle of each side of the square cross section and it is equal to the corresponding one observed in the case of the equivalent circular tube loaded under the same conditions, see Chapter 5. Therefore, the associated part of energy absorbed, which equals the external work, as can be obtained by measuring the area under the load/displacement curve in the elastic regime, see Figure 6.4, is

$$W_{Lc} = 2\cdot[\pi\cdot L_c\cdot(b-t)/2]\cdot R_{\mathrm{ad}} + n\cdot(t/2)\cdot G\cdot L_c = \int_0^{s_1} P\,ds = \frac{1}{2}P_{\max}\cdot s_1 \qquad (6.1)$$

where, following the Notation, R_{ad} is the fracture energy required to fracture a unit area of the adhesive at the interface between two adjacent layers, b the tube external side width, n the number of axial splits, G the fracture toughness and t the shell wall thickness.

The energy required for the deformation mechanism, regarding the history of the formation of the crush zone, see Chapter 5, equals the external work absorbed by the deforming shell in this regime, i.e.

$$W_{\mathrm{tr}} = [2\int_0^{\varphi}\sigma_{\mathrm{o}}\cdot l_s\cdot(l_s/2)d\varphi]\cdot 4\cdot(b-t) = \int_{s_1}^{s_2} P\,ds \qquad (6.2)$$

where, following the Notation, σ_o is the normal stress applied by the wedge to fronds, $l_s (= t/(2\cdot\sin\theta))$ the side length of the wedge inscribed to the bent fronds, φ $(=\alpha/2)$ the semi-angle of the wedge, see Figure 6.5(b), and s_2 is the related shell shortening corresponding to the completion of the wedge formation, see Figure 6.4.

Since the intrawall crack propagates at a constant speed equal to the speed of the crosshead of the press, it can be assumed that the crack length L_c remains constant. Also the length of the split of the crush zone (AB), at the contact side with the steel platen, approximates the wall thickness, t, see Figures 6.2(b) and 6.5(b). Therefore, taking into account the failure mechanism outlined above, the total dissipated energy for a crush distance, s can be estimated as follows:

- Energy dissipated due to friction between the annular wedge and fronds and between fronds and platen

$$W_{\text{i}} = 2 \cdot (\mu_{s1} \cdot P_1 + \mu_{s2} \cdot P_2) \cdot 4 \cdot (b - t) \cdot (s - s_2) \tag{6.3}$$

where, P_1 is the normal force per unit length applied by the platen to the internal and external fronds, P_2 is the normal force per unit length applied to the sides of the wedge, μ_{s1} is the coefficient of friction between frond and platen and μ_{s2} is the coefficient of friction between the wedge and the fronds. It must be noted that

$$P_2 = \sigma_{\text{o}} \cdot l_s \tag{6.4}$$

and

$$\sigma_{\text{o}} = k \cdot \sigma_{\theta} \tag{6.5}$$

where, k is a constant and σ_θ is the tensile fracture stress of the composite material.

- Energy dissipated due to fronds bending

$$W_{\text{ii}} = 2 \cdot [\int_0^{\varphi} P_2 \cdot (l_s/2) d\varphi + \int_{s_2}^{s} P_2 \cdot \varphi \cdot ds] \cdot 4 \cdot (b - t) \tag{6.6}$$

- Energy dissipated due to crack propagation

$$W_{\text{iii}} = R_{\text{ad}} \cdot [(s - s_1) \cdot 4 \cdot b + \pi \cdot L_c \cdot (b - t)] \tag{6.7}$$

- Energy dissipated due to axial splitting

$$W_{\text{iv}} = 4 \cdot (t/2) \cdot G \cdot s \tag{6.8}$$

From Equations (6.3), (6.4), (6.5) and (6.6) the total energy dissipated for the deformation of the shell can be calculated as

$$W_T = W_{\text{i}} + W_{\text{ii}} + W_{\text{iii}} + W_{\text{iv}} \tag{6.9}$$

therefore,

$$\begin{aligned} W_T = {} & [1/(1 - \mu_{s1} + \mu_{s1} \cdot s_2/s)] \cdot \{[4 \cdot (b/t) \cdot t \cdot k \cdot \sigma_\theta \cdot (s - s_2) \cdot [\mu_{s2}/\cos\varphi \\ & - \mu_{s1} \cdot (\tan\varphi + \mu_{s2})] + (\varphi/\cos\varphi) \cdot (025 \cdot t/\cos\varphi \\ & + s - s_2)] + R_{\text{ad}} \cdot (b - t) \cdot [4 \cdot (s - s_1) + \pi \cdot L_c] + 4 \cdot (t/2) \cdot G \cdot s\} \end{aligned} \tag{6.10}$$

The total normal force applied by the platen to the shell can be calculated as

$$P = W_T / s \tag{6.11}$$

The fracture tensile stess, σ_θ of the materials A and B can be estimated from the stress-strain curves shown in Figure 5.4 in Chapter 5. The static friction coefficients, μ_{s1} and μ_{s2} were obtained by employing the curling test [41]. The fracture toughness, G was estimated from the tension test of notched strips. The interfacial fracture energy, R_{ad} can be obtained from Equation (5.2), see Chapter 5, by employing the strip peel test [46].

THE EFFECT OF STRAIN-RATE

In Reference [84] some interesting remarks pertaining to the effect of the strain-rate on the crushing behaviour of axially loaded composite tubes are reviewed. Strain-rate and, therefore, crush speed, can influence the mechanical properties of the fibre and matrix. Matrix stiffness and failure strain can be a function of strain-rate. Therefore, the absorbed energy associated with the interlaminar crack growth, is affected by the crush speed. Note, however, that the mechanical properties of the brittle fibres, and the fracturing of the lamina bundles are, in general, insensitive to strain-rate. The coefficient of friction between the composite and crushing surface and between debris wedge and fronds is affected on the crush speed, and therefore, the energy-absorption capability of the shell can be influenced by changing it.

As outlined in Chapter 5, the major part of the energy absorbed during the static axial compression of a shell is dissipated as frictional work in the crushed material or at the interface between material and tool; this is estimated to be about 50% or more of the total work done. In dynamic collapse, attention is directed towards the influence of the strain-rate on the frictional work absorbed during the impact, taking into account all structural and material parameters, which may contribute to it, i.e. the fibre and matrix material, the fibre diameter and orientation in the laminate, the fibre volume content, as well as the conditions at the interfaces.

As mentioned above, two distinct regions, where the development of frictional forces is of great importance, may be identified, see Figures 6.2 and 6.5: the fronds/wedge contact region, composed of the same material, and the fronds/platen contact area, composed of different materials. The coefficient of friction in these regions depends on various phenomena associated with the material flow, such as: elastic and/or plastic deformation occurring at the contact area of the sliding surfaces subjected to external load; interfacial bonding due to the electrostatic forces developed in the contact area; adhesion occurring at the contact region during the sliding of the two deformable bodies of the same material. Note that interfacial bonding and adhesion result in an increase of the coefficient friction, whilst plastic deformation results in a decrease of the coefficient of friction due to the shear resistance at the sliding surfaces [105].

In general, impact loading results in high electrostatic forces, resulting, therefore, in higher values of the dynamic coefficient of friction. Note, however, that the elastic/plastic deformation imposed on the sliding interfaces and the associated surface changes are not clearly defined and it is, therefore, difficult to estimate the prevailing nature, static or dynamic, of the phenomenon occurring. Knowledge, therefore, of the conditions prevailing, which are not uniquely defined during the static or dynamic axial collapse, leads to an estimate of the mechanical response and the crashworthy behaviour of the structural component during the crushing process. The governing phenomena, outlined above, are greatly affected by the material properties, influencing finally the effectiveness of the collapsed component, as far as its crashworthy capacity is concerned.

The total energy absorbed during the dynamic axial loading of square tubes can be, therefore, estimated from Equation (6.10) by substituting only the friction coefficients, μ_{s1} and μ_{s2} with the dynamic ones, μ_{d1} and μ_{d2}, respectively, as

$$\begin{aligned} W_T = {} & [1/(1-\mu_{d1}+\mu_{d1}\cdot s_2/s)]\cdot\{[4\cdot(b-t)\cdot t\cdot k\cdot\sigma_\theta\cdot(s-s_2)\cdot[\mu_{d2}/\cos\varphi \\ & -\mu_{d1}\cdot(\tan\varphi+\mu_{d2})]+(\varphi/\cos\varphi)\cdot(0.25\cdot t/\cos\varphi \\ & +s-s_2)]+R_{ad}\cdot(b-t)\cdot[4\cdot(s-s_1)+\pi\cdot L_c]+4\cdot(t/2)\cdot G\cdot s\} \end{aligned} \tag{6.12}$$

It is assumed that the tensile fracture stress and the fracture toughness of the material remain constant, although these parameters may be a function of the strain-rate.

An estimate of the dynamic friction coefficients, μ_{d1} and μ_{d2} was obtained by comparing the experimental crushing loads in Table 6.1.

6.3.5 Crashworthy Capability: Concluding Remarks

In the previous Sections of the present chapter, the fracture behaviour and the crashworthy characteristics of composite square tubes subjected to axial static and dynamic loading is examined in detail.

Two collapse modes were observed; the stable progressive collapse mode, Mode I, associated with large amounts of crush energy, resulting, therefore, in a high crashworthy capacity of the structural component and, the mid-length collapse mode, Mode III, featuring a brittle fracture involving catastrophic failure. Regarding the microfracture mechanism of square tubes subjected to axial loading, as far as the Mode I of collapse is concerned, the experimental observations revealed that square tubes developed almost the same deformation mechanism to that observed for circular tubes, with the difference that the crush zone changes in size over the tube circumference. In general, this mechanism applies to both statically and dynamically loaded shells, the only differences encountered with the shape of the wedge and the microcracking development.

The post-crushing regions of the load/displacement curves for the statically and dynamically loaded shells, which followed the Mode I of collapse, revealed distinct differences; the dynamically obtained curves are highly serrated, probably due to the impact nature of the loading and the inertia response of the load cell, whilst the static curves show the formation of typical peaks and valleys with very small fluctuations.

For the square tubes made of material B, it is evident that the mean post-crushing load and the energy absorbed are mainly affected by the crushing length, whilst the axial length of the shell has no significant effect on these crashworthy characteristics. Dynamic collapse overestimates static collapse by about 20%, as far as the crashworthy characteristics is concerned, probably due to higher values of the dynamic friction coeffiefficients between the wedge/fronds and the fronds/platen interfaces. Contrariwise, for square tubes made of material A and subjected to dynamic testing, lower values by about 15% were obtained than those predicted in static testing, probably due to the different values of the dynamic friction coefficients of the two materials A and B (see above), leading, therefore, to the conclusion that the characteristics of the shell material determine the crashworthy behaviour of the impacted shells.

The mean post-crushing load, $\overline{P}$, the energy absorbed, W, and the specific energy, W_s were well predicted theoretically by the proposed analysis, within $\pm 12\%$ for both materials tested, see Table 6.1. Note that in this approach fibre orientation was not taken into account. According to the proposed theoretical analysis, the distribution of the dissipated energy of the collapsed shell, associated to the four main energy sources, was estimated as:

- Energy due to friction between the annular wedge and the fronds and between the fronds and the platen, estimated to about 44% and 48% of the total one for the materials A and B, respectively
- Energy due to fronds bending, about 48% and 32%, respectively
- Energy due to crack propagation, about 7% and 18%, respectively
- Energy due to axial splitting at the four corners of the shell, about 1% and 2%, respectively

Note that the contribution of the frictional conditions between wedge/fronds and fronds/platen to the energy absorbing capability is more significant than the other ones. As discussed above, it mainly depends upon the friction coefficients μ_{s1}, μ_{d1} and μ_{s2}, μ_{d2}, which are affected by the surface conditions at the interfaces composite material/platen or drop mass and composite material/debris wedge, respectively. From the analysis, it is also evident that the annular debris wedge supports the 55% of the crushing load and the internal and external fronds the remaining 45% for both materials tested.

It was also observed that the energy absorbed due to bending of the fronds is greatly affected by the magnitude of the wedge semi-angle, which varies between 45° and $\varphi = \tan^{-1}(1/\mu_2)$.

In general, from the energy absorbing capacity point of view, square tubes of material B seem to be more efficient, as predicted both theoretically and experimentally.

6.4 BENDING

6.4.1 Experimental

Bending crush tests were carried out to observe the modes of collapse and the crashworthy characteristics of thin-walled composite components; they were performed on cantilevered beams of square or rectangular sections, bent about their weak or strong axis.

The experimental set-up, described in detail in Chapter 5, is schematically shown in Figure 6.6. It is equipped with a complex device providing the continuous meas-

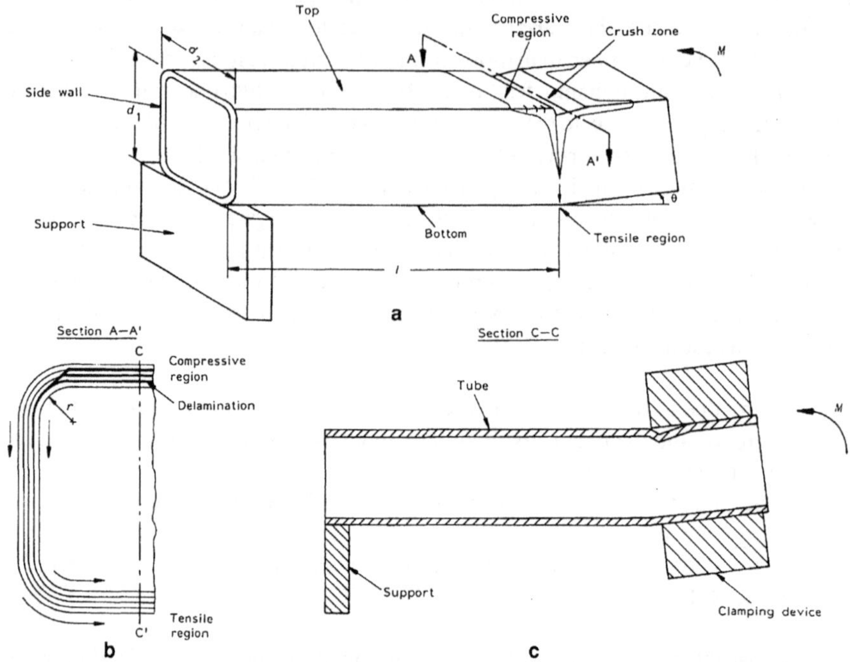

Figure 6.6. (a) Cantilever thin-walled tube under bending, (b) cross section at the crush zone, (c) longitudinal section of the deformed tube.

urement of the effective bending moment and the angle of rotation as the member collapses. This approach uses a deformation (angle of rotation) controlled loading to obtain the diagram as a function of angle of rotation. The moment applied to the tubular beam is measured directly by a torque cell, attached to the speed reducer shaft, whilst the hinge rotation was measured using a linear continuous carbon potentiometer, attached to the end of the speed reducer output shaft, as shown in Figure 5.8 of Chapter 5. The transducer output is connected to a PC-based data-acquisition, using Labtech Notebook software to collect the data, as the tests were conducted. The variation of the bending moment with hinge rotation for each loaded specimen is directly obtained from this system.

For the bending tests, the same materials used for the axial loading tests, e.g. the fibreglass composite materials A and B, were employed. For more details about laminate fibre lay-up and the fabrication of the composite tubes, see Section 5.3.1 In Chapter 5.

Experimental results pertaining to the collapse modes observed, the loading and energy absorbing characteristics, as well as the dimensions and lay-up of the specimens tested, are tabulated in Table 6.2. Photographs showing characteristic terminal views of the deformed specimens, with the various collapse modes at macroscopic scale, are presented in Figures 6.7–6.11.

Table 6.2: Crushing characteristics of square and rectangular tubes subjected to bending

Specimen no	Material	d_1 (mm)	d_2 (mm)	d_2/d_1	No of layers	Thickness t (mm)	θmax (°)	Peak moment, M_{max} (N m)		Energy absorbed W_T(J)
								Experimental	Theoretical	
1	A	40.0	40.0	1.0	1	1.5	1.4	22.6		0.59
2	A	40.0	40.0	1.0	2	2.4	2.5	59.8	56.8	1.43
3	A	40.0	40.0	1.0	4	4.8	22.0	610.2	598.8	140.50
4	A	50.5	40.0	0.8	1	1.9	4.5	67.8		2.66
5	A	50.6	41.0	0.8	2	2.7	20.0	265.1	250.2	56.50
6	A	51.8	40.5	0.8	3	3.9	27.0	435.0	401.7	105.32
7	A	51.2	40.5	0.8	4	4.8	24.0	802.2	768.6	194.43
8	A	40.0	50.5	1.25	2	2.8	3.8	127.4	118.5	4.26
9	A	78.0	40.0	0.5	1	1.5	31.0	77.5		43.27
10	A	76.3	40.9	0.5	2	2.6	26.0	367.2	358.9	95.88
11	A	79.7	40.4	0.5	3	3.8	23.0	790.9	746.3	186.12
12	A	77.8	40.3	0.5	4	4.9	28.0	1231.5	1158.7	279.18
13	A	d=52.6*			3	4.0	14.0	562.6		91.56
14	B	56.0	56.0	1.0	**	2.8	11.7	580.4	567.0	106.01
15	B	56.0	56.0	1.0	**	2.8	9.6	549.0	567.0	112.80

* Specimen 13 represents the equivalent circular specimen to specimen 11

** See Section 5.3.1 for material B lay-up

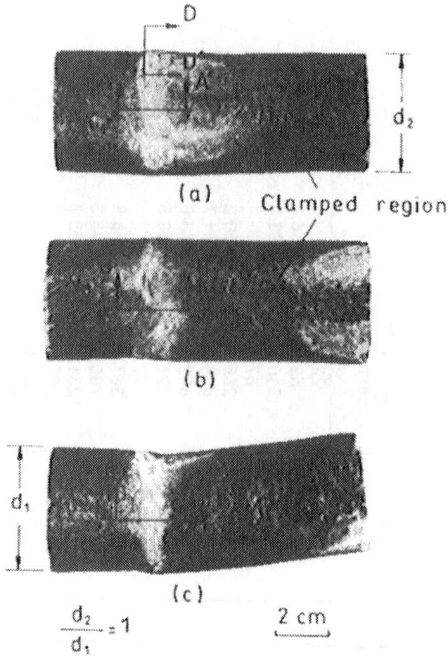

Figure 6.7. Macroscopic views of specimen 3 (see Table 6.2); (a) compression zone, (b) tension zone, (c) side wall.

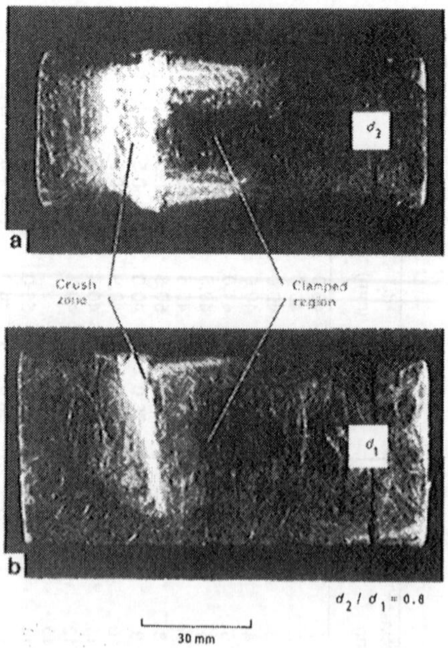

Figure 6.8. Macroscopic views of specimen 7 (see Table 6.2); (a) compression zone, (b) side wall.

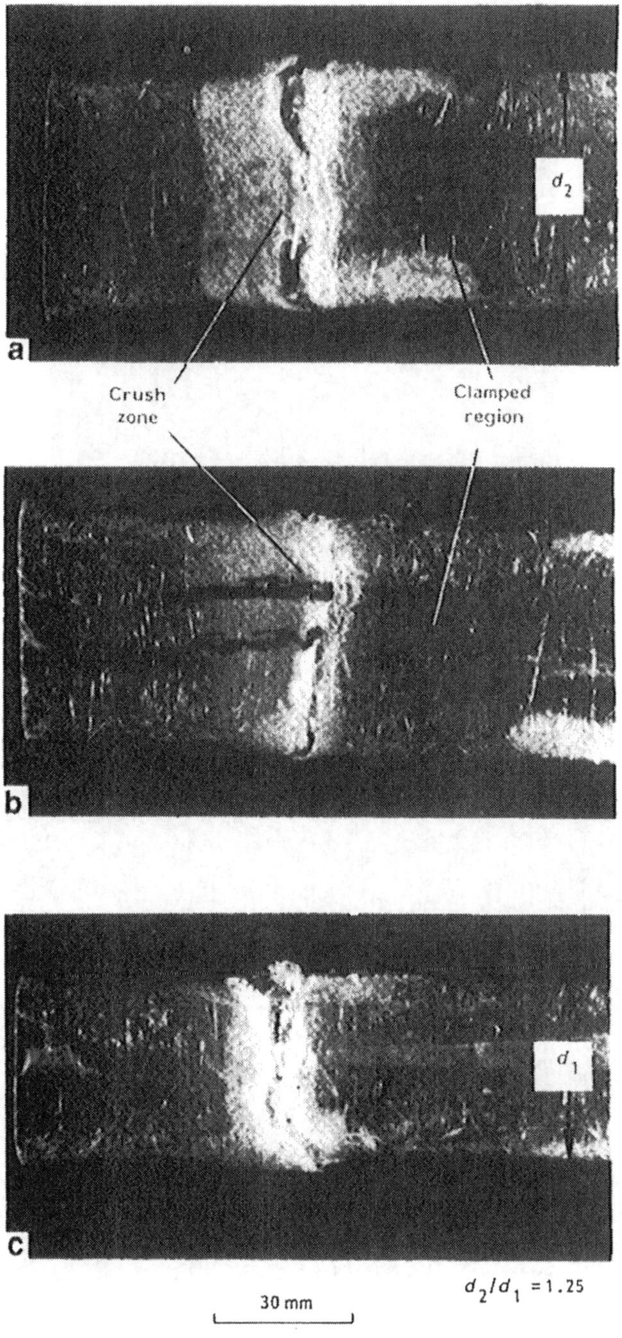

Figure 6.9. Macroscopic views of specimen 8 (see Table 6.2); (a) compression zone, (b) tension zone, (c) side wall.

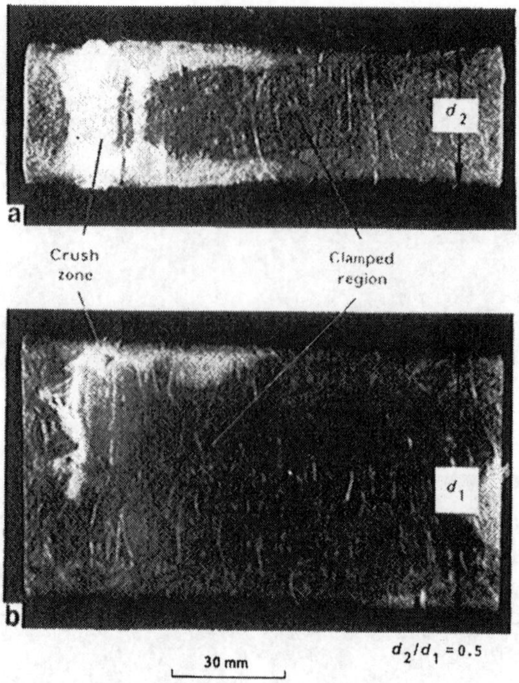

Figure 6.10. Macroscopic views of specimen 10 (see Table 6.2); (a) compression zone, (b) side wall.

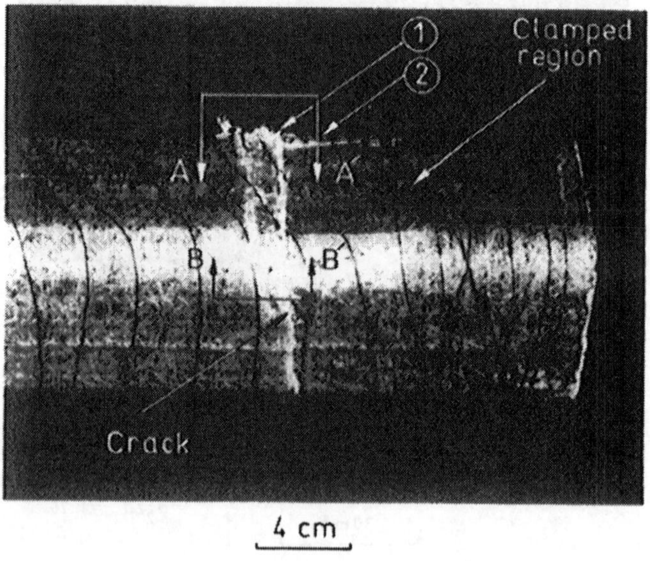

Figure 6.11. Microscopic side view of specimen 14 (Material B; see Table 6.2) after collapse.

To prepare the metallographic specimens, strips were cut from the damaged region and removed from the shell wall. The relevant surface was then prepared and polished properly; see Section 5.4.1 in Chapter 5. The main microfractures, developed in the various crushed regions, were obtained by using a UNIMET metallographic optical microscope. Micrographs of the main fractured and collapsed areas of the tube are presented in Figures 6.12–6.14.

The values of the maximum bending moment, M_{max} and the energy absorbed, W, obtained by carefully measuring the area under the bending moment-angle of rotation curves shown in Figures 6.15–6.19, are also tabulated in Table 6.2.

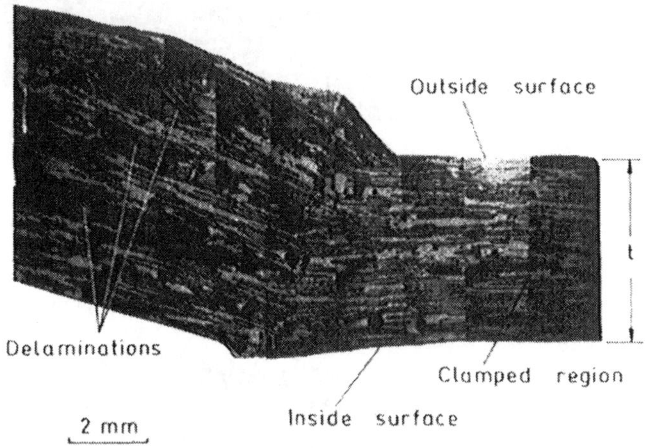

(a)

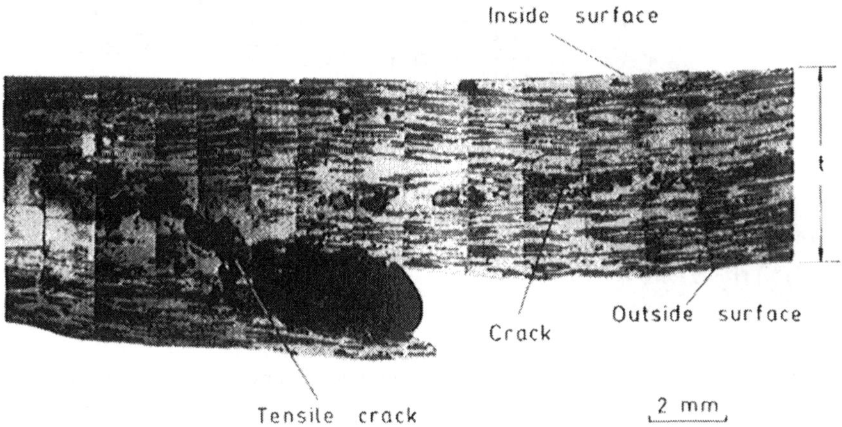

(b)

Figure 6.12. Micrographs showing microfailures observed at various crash zones of specimen 3 (see Table 6.1. and also Figure 6.7.); (a) section AA′ (compression zone), (b) section BB′ (tensile zone).

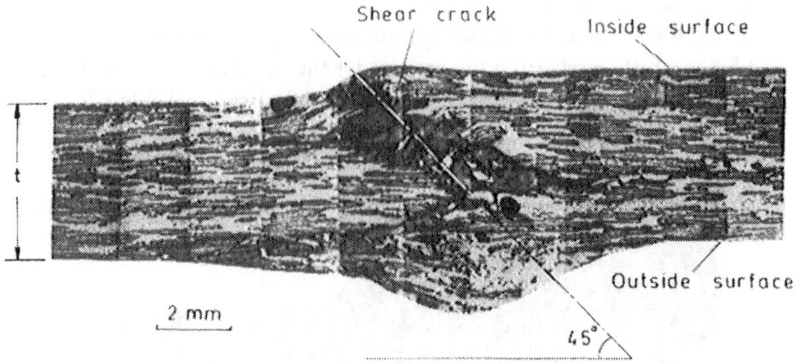

(c)

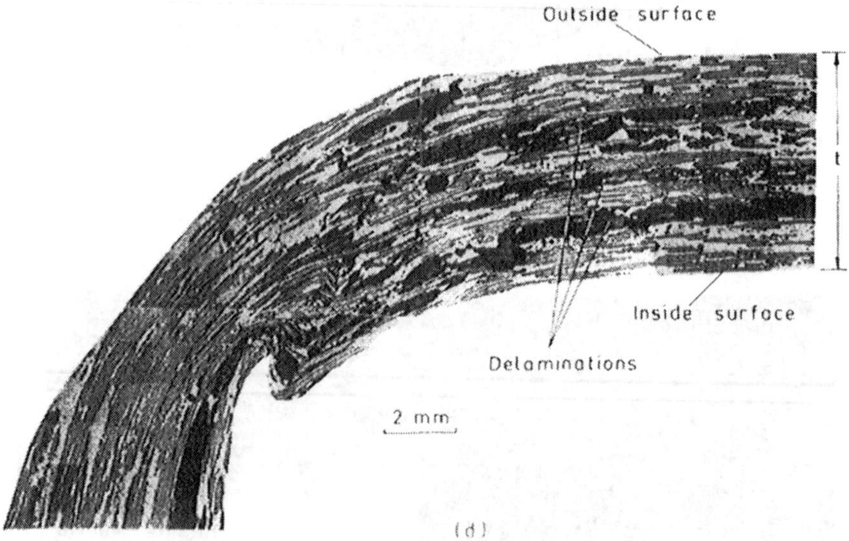

(d)

Figure 6.12 (continued). Micrographs showing microfailures observed at various crash zones of specimen 3 (see Table 6.1. and also Figure 6.7.); (c) section CC′ (side wall), (d) section DD′ (corner).

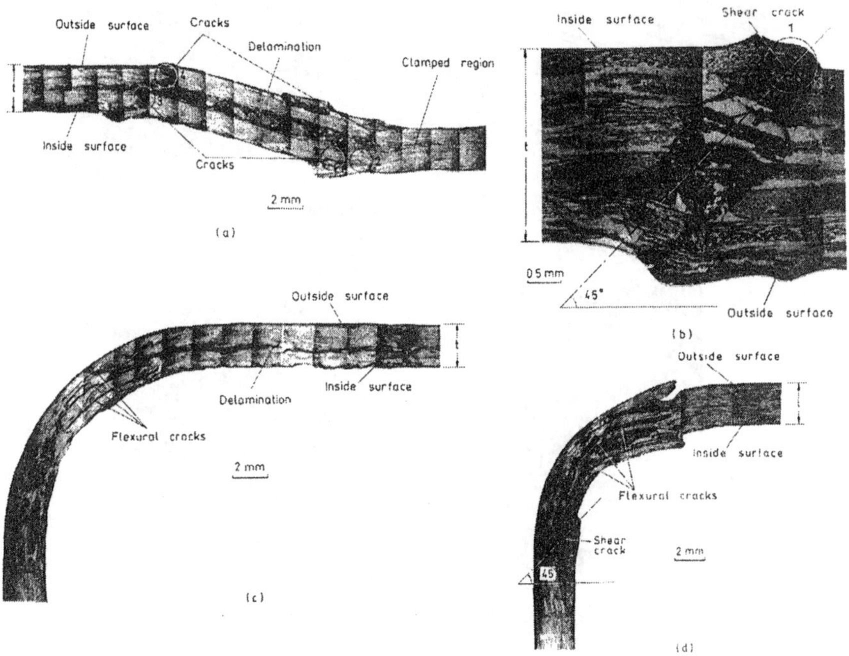

Figure 6.13. Micrographs showing microfailures observed at various crash zones of specimen 14 (see Table 6.1.1); (a) section AA′ (compression zone), (b) section BB′ (tensile zone), (c) position 1 showing delamination cracking near the corner, (d) position 2 showing flexural cracking near the corner.

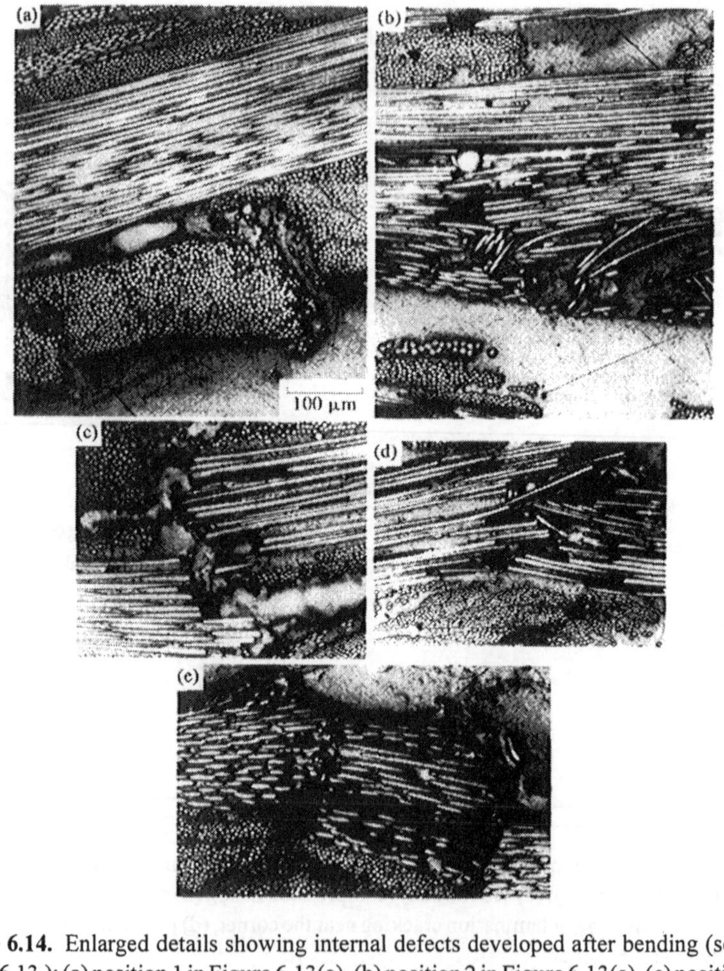

Figure 6.14. Enlarged details showing internal defects developed after bending (see also Figure 6.13.); (a) position 1 in Figure 6.13(a), (b) position 2 in Figure 6.13(a), (c) position 3 in Figure 6.13(a), (d) position 4 in Figure 6.13(a), (e) position 1 in Figure 6.13(b).

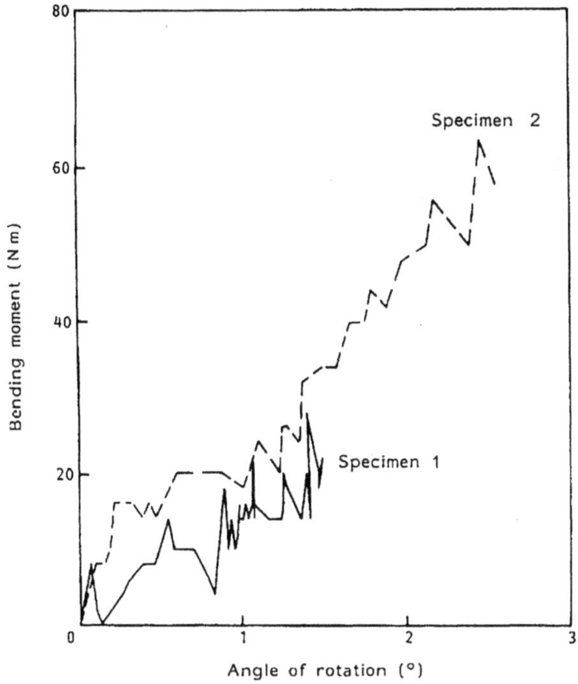

Figure 6.15. Bending moment/angle of rotation characteristics of bent tubes with $d_2/d_1 = 1$ (see Table 6.2 for details).

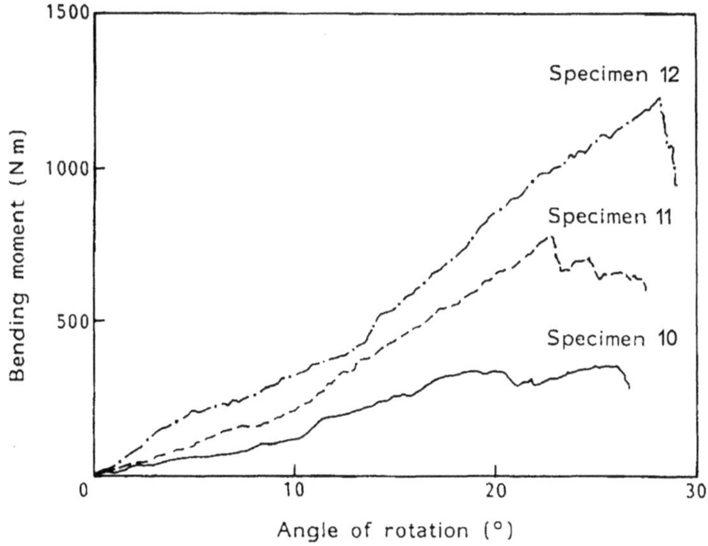

Figure 6.16. Bending moment/angle of rotation characteristics of bent tubes with $d_2/d_1 = 0.5$ (see Table 6.2 for details).

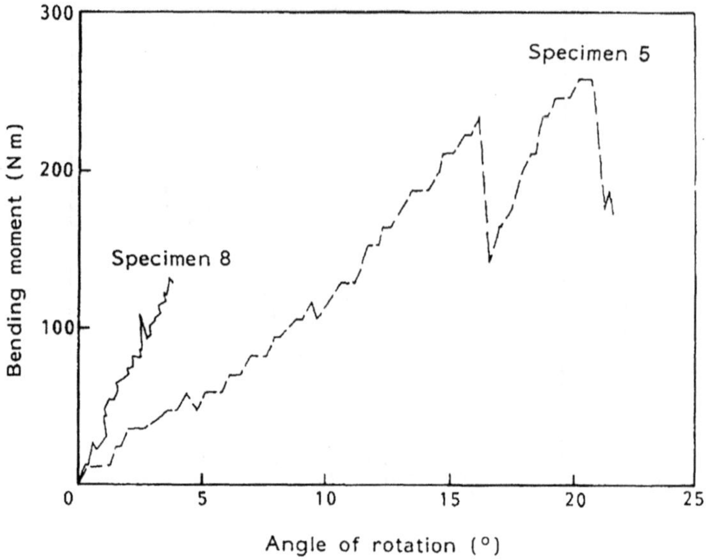

Figure 6.17. Bending moment/angle of rotation characteristics of bent tubes (see Table 6.2 for details).

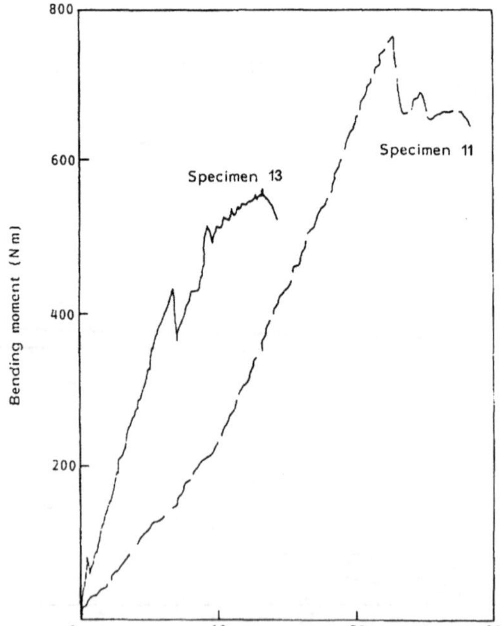

Figure 6.18. Bending moment/angle of rotation characteristics for equivalent-sectioned circular and rectangular specimens (see Table 6.2 for details).

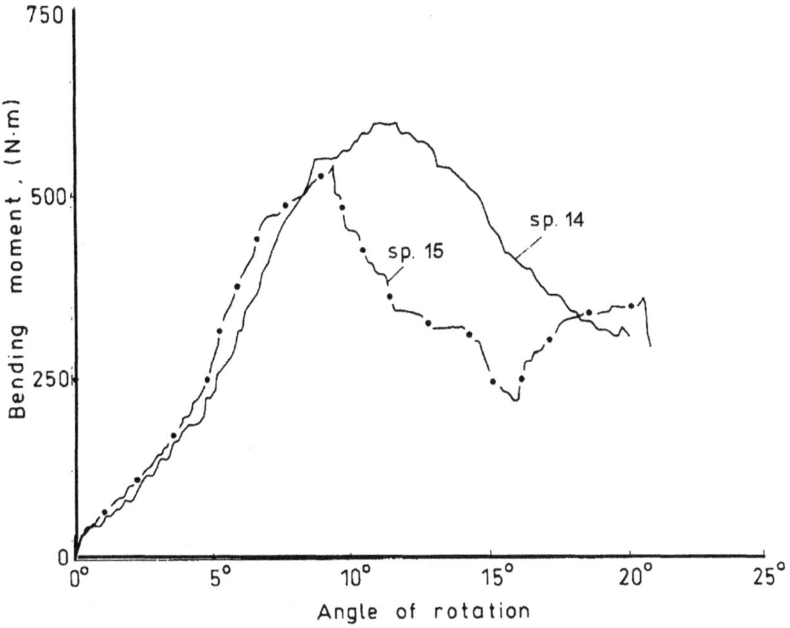

Figure 6.19. Bending moment/angle of rotation characteristics for tubes of material B.

6.4.2 Failure Mechanisms: Experimental Observations

DEFORMATION MODES

Three distinct regions with different macro- and microscopic characteristics were observed throughout the bending process of rectangular tubes. The top side of the tube, see Figure 6.6, is predominantly subjected to compression, the bottom one to extensive tensile straining, whilst the side walls of the tube are associated with combined compression/tension features. The four corners of the rectangular cross section greatly influence the above-mentioned mechanism, leading to the deformation modes shown in Figures 6.7–6.11.

Experimental macroscopic observations indicate that the predominant failure mode is extensive microfracturing of the tubular component, adjacent to the edge of the clamping device in the compression zone, see Figure 6.6; this is probably due to the maximum bending moment attained there.

Fracturing initiates on the top side of the tube adjacent to the clamping edge, which is subjected to compression, and following a typical progressive failure mechanism as the hinge develops, the fracture region spreads quickly from the centre of the top side towards the corners; a large number of cracks/delaminations develop at the corners, just underneath the clamping edge. As the hinge rotation continues, the tube sustains higher loads as the corners begin to crush, whilst the compressive top side and the side walls begin to buckle. After the maximum strength was attained, the compressive top side and the side walls failed due to buckling and/or splintering, leading finally to an overall separation of the tube along the compressive region and the side walls. The extent of tube separation is influenced by the wall thickness, being more profound for thinner specimens.

The main, therefore, features of the energy absorbing mechanisms observed during bending, shown schematically in Figure 6.6 and in the photographs of Figures 6.7–6.11 and outlined above, are: extensive microfracturing of the compressive side of the tubes; buckling and crushing of the side walls accompanied by delaminated regions extended over a wide region of the corners of the tube (white areas in Figures 6.7–6.11); simultaneous shifting of the neutral axis towards the tensile side, leading finally to deep collapse and separation of the tube wall.

MICROSCOPIC OBSERVATIONS

From the terminal deformation patterns of the bent specimens, shown in Figures 6.12–6.14, the following microscopic observations may be made for each of the above-mentioned deformation zones.

(i) Top Compression Zone

In the compressive region a tendency for local buckling is apparent; bending depends on the fibre properties and the support given by the matrix of the composite material. Compressive failure is characterised by slipping and/or fracturing of the matrix material, see Figures 6.12(a), 6.13(a) and 6.14.

For very thin, one-layered, tubular components, the microbuckling phenomena are accompanied by the development of internal cracks due to the combined action of shearing and flexural straining. Shell separation was not observed for such specimens. However, for multi-layered specimens, the delamination mechanism is usually activated, see Figures 6.12(a) and 6.13(a). Internal small cracks, perpendicular to the longitudinal axis of symmetry of the tube, are found close to the region of maximum deformation of one- or two-layered specimens, although they are not visible macroscopically. Similar cracks of random orientation, mainly due to material fragmentation, were also observed in three- and/or four-layered tubes.

Separation of the shell wall occurs at an angle of about 45° to its longitudinal axis of symmetry following the direction of maximum shear stress.

A usually formed central delamination crack propagates longitudinally through the wall thickness, being bifurcated to smaller ones and crossing regions with small fibre volume fraction, or through the boundaries between fibres of different orientation.

(ii) Bottom Tensile Zone

The bottom tensile region is characterised by a "brush-like" appearance of pulled-out fibres over the separation area of the shell. The failure mechanism, governing the tensile loading conditions, results in significant straining perpendicular to the fibres, causing transverse cracking of heterogeneous nature, including matrix cracking, fibre/matrix interface debonding of transverse fibres, delamination and fibre breakage of longitudinal ones; see Figure 6.12(b).

(iii) Side Walls

From the failure surface topography of the side walls of the tube, see Figures 6.12(c), 6.13(b) and 6.14(e), a complex damage mechanism can be suggested, consisting of the combination of the following straining conditions:

- Longitudinal tensile failure, causing cracking normal to the fibres, probably due to the simultaneous action of the following collapse phenomena: (a) elastic straining of the brittle matrix and of the fibres leading to crack propagation with small amount of the energy absorbed; (b) debonding of fibres from the resin, mainly observed in broken areas of white colour; (c) pull-out of broken fibres from their sockets in the resin, usually after debonding occurred.

  Factors affecting these failure phenomena are the interfacial bond strength, the failure strength of the fibres and the matrix, the volume fraction, the diameter of the fibres and the moduli of the fibres and the matrix.
- Longitudinal compressive failure, with the compressive strength strongly depended on factors, such as: fibre and resin properties; interface strength and void content; fibre volume fraction. Forseen secondary collapse mechanisms superimposed in this case are: (a) elastic local microbuckling of the composite material, mainly affected by the elastic properties of the material. It is detected as areas with low volume fractions, where out of phase fibres buckling developed, or as areas with higher volume fractions with in-phase buckling, probably due to localised fracture of the fibres and the fibre-resin interface; (b) shear failure of resin and fibres through a kink band, progressing across the specimens. Misalignment of the fibres in relation to the loading axis, microstructural inhomogeneities, voids and buckling of the fibres are some of the factors causing this kind of failure.
- Transverse tensile failure, inducing cracking in the longitudinal direction. This type of collapse is also governed and influenced by several factors, such as: matrix and fibres strength, interface bonding, presence and distribution of voids; internal stresses and strain distributions due to stacking conditions, mainly over the corner area, inhomogeneities within the matrix etc.

  Weak interfacial bonding leads to significant reduction of the strength of the composite material since the fibres contribute very little to the strength. On the contrary, with fibres and matrix strongly bonded, the strength of the composite material in the transverse direction probably depends on the bond strength and the matrix properties. In this case, the mechanism of crack propagation develops in the form of matrix cracking and/or debonding at the fibre-matrix interface accompanied by an amount of fibre pull-out. In general, transverse cracks initiate in and propagate through regions of dense fibre packing, whereas they are arrested in resin-rich regions.

- Transverse compression failure, leading to collapse under a shear mechanism, depended on the orientation of the fibres in the plane of maximum shear stress.
- Pure shear failure by forces acting parallel to the fibres. The strength is dominated by the matrix properties, as crack propagation occurs through the matrix disturbing or breaking the fibres. The shear strength is affected by the strength of the interfacial bond and the stress concentration effects, associated with the presence of fibres and voids formed during the stacking sequence.

The shear failure may also develop as inter- and intralaminar shear without disturbing the fibres. During intralaminar shear, some misaligned fibres tend to hold the crack together, causing in this manner an increase of the fracture toughness. With interlaminar shear, the cracks developed can propagate entirely through the resin between the laminae.

(iv) Corners

The regions close to the corners are subjected to complex straining, leading to uncontrolled fracture patterns, see Figure 6.12(d), 6.13(c) and (d), accompanied by severe delamination cracks of the neighbouring areas, mainly affecting the buckling behaviour of the compression zone of the tube.

6.4.3 Energy Absorbing Characteristics

The following remarks may be drawn from the obtained M/θ graphs, see Figures 6.15–6.19:

- Initially the tube behaves elastically. A steady state small increase of the slope of the bending moment corresponds to elastic bending, accompanied by extensive microfracturing of the tube wall in the compressive region. Then, the slope of the M/θ curve changes with increasing bending moment, up to a peak value M_{max}. During this phase, the load is carried by the compressive top side and the side walls of the tube, due to buckling and crushing in the extended region of the hinge rotation. As the maximum moment is attained, after a considerable amount of hinge rotation, the edges undergo severe material fragmentation, leading to a decrease of the bending moment. Deep collapse follows, where the fractured edges interact with each other, causing further resistance to the bending of the tube.
- The maximum bending moment seems to mainly depend upon the strength of the compressive top side and the buckling strength of the tube side walls.
- The post-crushing region (deep collapse) usually is greatly serrated and depends on the extension and the mode of collapse following the first cracking; compare the curves of specimens 14 and 15 in Figure 6.18.
- The energy absorbing efficiency of the tube increases with increasing wall thickness, see Figures 6.15–6.19. For thicker specimens (four-layered), however, the maximum rotation angle sharply diminishes, resulting in small

amounts of absorbed energy; this is probably due to severe local fragmentation of the composite material at the top compressive zone.

- The peak bending moment increases with increasing wall thickness, see Figures 6.15–6.19. It seems also to increase with decreasing ratio d_2/d_1, see Notation. It must be noted that specimens with $d_2/d_1=1$ and 1.25 show low deformation capacity, undergoing small angles of rotation and developing poor energy absorption efficiency; see Figure 6.17 and Table 6.1.
- From the comparison of specimen 11 with its equivalent circular tube, see Figure 6.18, it can be seen that the circular tube initially behaves in a better manner, as far as absorption efficiency is concerned. As the deformation progresses, the corners of the rectangular tube play a significant role by improving the absorption efficiency, leading, therefore, to better crashworthiness characteristics for the rectangular tubular components.

6.4.4 Failure Analysis

The classical lamination theory is suitably modified to cover the deformation mechanism of a curved laminate, consisting of a number of laminae stacked together at various planar orientations and loaded in a prescribed manner; see Figure 6.20.

The deformation mechanism of a loaded laminate is subjected to the following assumptions:

- Lines, perpendicular to the undeformed circumferential axis of the laminate, remain lines and perpendicular to the deformed circumferential axis.
- Lines, perpendicular to the undeformed longitudinal axis of the laminate, remain lines and perpendicular to the deformed longitudinal axis.
- The plies are linear elastic and homogeneous orthotropic.

The displacements in x and y directions see Figures 6.21 and 6.22, may be expressed respectively as

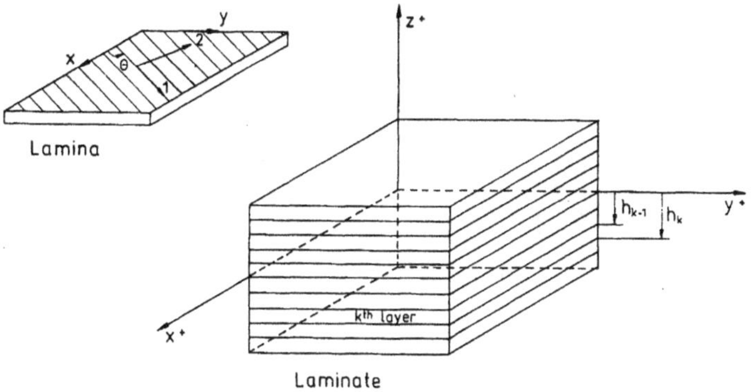

Figure 6.20. Configuration of a laminated plate.

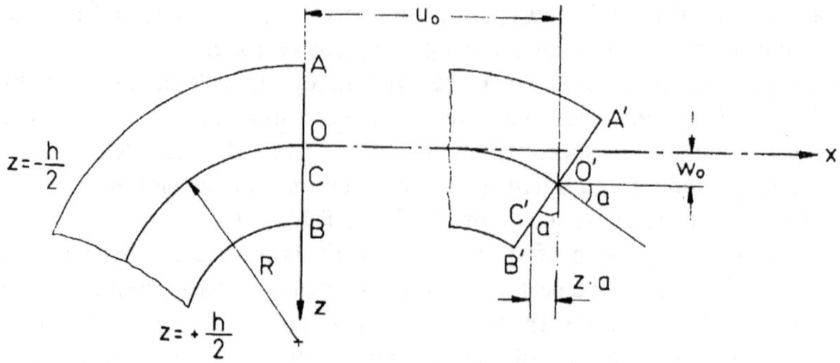

Figure 6.21. Bending of line element in *x-z* plane.

$$u = u_o - z\alpha = u_o - z\frac{\partial w_o}{\partial x} \tag{6.13}$$

$$v = v_o - z\beta = v_o - z\frac{\partial w_o}{\partial y} \tag{6.14}$$

where, *x,y,z* is the global laminate coordinate system.

The normal strain is equal to:

$$\varepsilon_x = \frac{\partial u}{\partial x_z} \tag{6.15}$$

where, the suffix *z* denotes that the initial unit length varies with z, see also Figure 6.22. From Figure 6.23, it is obtained,

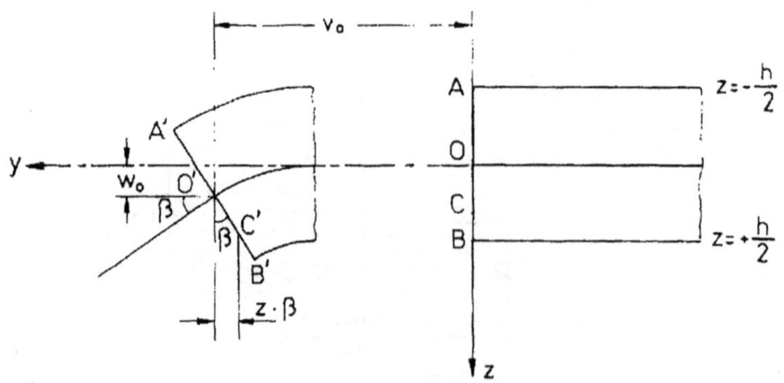

Figure 6.22. Bending of line element in *y-z* plane.

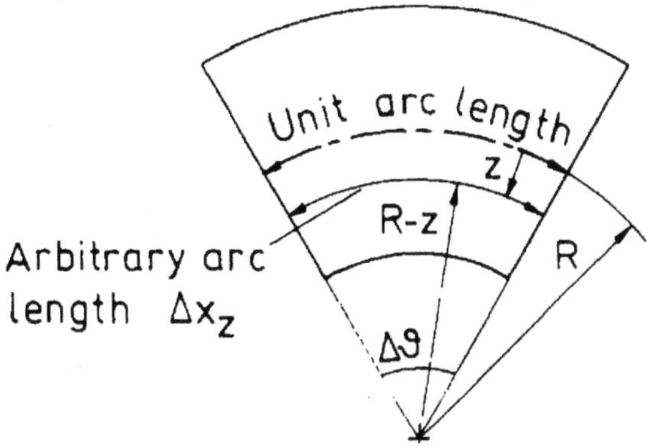

Figure 6.23. Differential arc length in *x-z* plane.

$$\frac{\Delta x}{\Delta x_z} = \frac{\Delta\theta \cdot R}{\Delta\theta \cdot (R - z)} = \frac{1}{1 - z/R} \tag{6.16}$$

Taking also into account that

$$\lim_{\Delta_x \to 0} \frac{\Delta x}{\Delta x_z} = \frac{\partial x}{\partial x_z} = \frac{1}{1 - z/R} \tag{6.17}$$

and, re-arranging Equation (6.15) by using the chain differentiation rule in the form

$$\varepsilon_x = \frac{\partial u}{\partial x_z} = \frac{\partial u}{\partial x}\frac{\partial x}{\partial x_z} \tag{6.18}$$

it finally yields,

$$\varepsilon_x = \left(\frac{\partial u_o}{\partial x} - z\frac{\partial^2 w_o}{\partial x^2}\right)\frac{1}{1 - z/R} \tag{6.19}$$

Similarly, it can be shown that for the axial strain, ε_y and the shear strain, γ_{xy} the following expressions can be obtained, respectively:

$$\varepsilon_x = \frac{\partial v_o}{\partial y} - z\frac{\partial^2 w_o}{\partial y^2} \tag{6.20}$$

$$\gamma_{xy} = 2\varepsilon_{xy} = \frac{\partial u}{\partial y} + \frac{\partial v}{\partial x_z} = \frac{\partial u_o}{\partial y} + \frac{\partial v_o}{\partial x}\frac{1}{1 - z/R} + z\left\{\frac{\partial^2 w_o}{\partial x \partial y}\left(1 + \frac{1}{1 - z/R}\right)\right\} \tag{6.21}$$

The above strain-displacement relationships, Equations (6.19), (6.20) and (6.21), rewritten in terms of the mid-surface strains, $[\varepsilon^o]$ and the mid-surface curvatures, $[k]$, result in

$$\begin{bmatrix} \varepsilon_x \\ \varepsilon_y \\ \gamma_{xy} \end{bmatrix} = \begin{bmatrix} \varepsilon_x^o \dfrac{1}{1-z/R} \\ \varepsilon_y^o \\ \dfrac{\partial u_o}{\partial y} + \dfrac{\partial v_o}{\partial x} \dfrac{1}{1-z/R} \end{bmatrix} + z \begin{bmatrix} k_x \dfrac{1}{1-z/R} \\ k_y \\ -\dfrac{\partial^2 w_o}{\partial x \partial y}\left(1+\dfrac{1}{1-z/R}\right) \end{bmatrix} \tag{6.22}$$

where,

$$\left.\begin{aligned} \varepsilon_x^0 &= \frac{\partial u_o}{\partial x}, \quad & \varepsilon_y^o &= \frac{\partial v_o}{\partial y} \\ k_x &= -\frac{\partial^2 w_o}{\partial x^2}, \quad & k_y &= -\frac{\partial^2 w_o}{\partial y^2} \end{aligned}\right\} \tag{6.23}$$

As R approaches infinity, in the case of a flat plate, Equation (6.22) reduces to that produced by the lamination theory [49].

From the constitutive relations for a lamina composed of a generally orthotropic material, relations, for a laminate consisting of several laminae bonded together, can be generated, see Figure 6.20 and Reference [49]. Thus, for the ith lamina of the laminate it can be written:

$$[\sigma]_i = [\overline{Q}]_i [\varepsilon] \tag{6.24}$$

or transforming it to the (1,2,3) principal material coordinate system, the following expression can be obtained

$$\begin{bmatrix} \sigma_1 \\ \sigma_2 \\ \tau_{12} \end{bmatrix} = [T] \begin{bmatrix} \sigma_x \\ \sigma_y \\ \tau_{xy} \end{bmatrix} \text{ and } \begin{bmatrix} \varepsilon_1 \\ \varepsilon_2 \\ \varepsilon_{12} \end{bmatrix} = [T] \begin{bmatrix} \varepsilon_x \\ \varepsilon_y \\ \varepsilon_{xy} \end{bmatrix} \tag{6.25}$$

and, finally

$$\begin{bmatrix} \sigma_1 \\ \sigma_2 \\ \tau_{12} \end{bmatrix} = [Q] \begin{bmatrix} \varepsilon_x \\ \varepsilon_y \\ \varepsilon_{xy} \end{bmatrix} \tag{6.26}$$

where,

$$[\sigma]=\begin{bmatrix}\sigma_x\\ \sigma_y\\ \tau_{xy}\end{bmatrix}$$

is the stress tensor referring to the global laminate coordinate system, and

$$[\varepsilon]=\begin{bmatrix}\varepsilon_x\\ \varepsilon_y\\ \varepsilon_{xy}\end{bmatrix}$$

is the strain tensor as obtained from Equation (6.22) above.

$$[T]=\begin{bmatrix}m^2 & n^2 & 2mn\\ n^2 & m^2 & -2mn\\ -mn & mn & (m^2-n^2)\end{bmatrix}$$

where, $m = \cos\theta^*$, $n = \sin\theta^*$ (θ^* is the angle between axis 1 and x), and

$$[Q]=\begin{bmatrix}Q_{11} & Q_{12} & 0\\ Q_{21} & Q_{22} & 0\\ 0 & 0 & 2Q_{66}\end{bmatrix}=\begin{bmatrix}\dfrac{E_{11}}{1-v_{12}v_{21}} & \dfrac{v_{12}E_{22}}{1-v_{21}v_{12}} & 0\\ \dfrac{v_{21}E_{11}}{1-v_{12}v_{21}} & \dfrac{E_{22}}{1-v_{12}v_{21}} & 0\\ 0 & 0 & \dfrac{E_{11}}{1+v_{12}}\end{bmatrix}$$

$[\overline{Q}]=[T]^{-1}[Q][T]$ is the generally orthotropic lamina stiffness.

For the prediction of the ultimate strength of the laminate under complex straining, each lamina must satisfy an appropriate failure criterion before full rupture. The 2-dimensional Tsai-Wu failure criterion [52] is used in the present case, expressed in the form

$$F_{11}\sigma_1^2+F_{22}\sigma_2^2+2F_{12}\sigma_1\sigma_2+F_{66}\sigma_6^2+F_1\sigma_1+F_2\sigma_2+F_6\sigma_6=1 \qquad (6.27)$$

where,

$F_{11}=1/X_tX_c$
$F_{22}=1/Y_tY_c$
$F_{66}=1/S^2$
$F_1=1/X_t-1/X_c$
$F_2=1/Y_t-1/Y_c$

$$\sigma_6 = \tau_{xy}$$
$$F_6 = F_{12} = 0$$

X_t, Y_t are the uniaxial tensile strengths of the lamina in the longitudinal and transverse direction, respectively, and X_c, Y_c the related uniaxial compressive ones; S is the shear strength of the lamina.

For the cantilevered laminated composite tube of Figure 6.24, subjected to an end shear load P, the vertical deflection at the loaded end may be calculated from:

$$w = \frac{PL^3}{3E_{XX}I_{XX}} \tag{6.28}$$

where, E_{XX} is the modulus of elasticity in the X-direction and I_{XX} the moment of inertia of the tube cross section. If X – Y is the coordinate system applied to the beam, then, according to the simple theory of flexure, the state of stress in the compression and tension zones for lamina i may be expressed as:

$$\begin{bmatrix} \sigma_y \\ \sigma_y \\ \tau_{xy} \end{bmatrix} = \begin{bmatrix} \pm PLY^i / I_{XX} \\ 0 \\ 0 \end{bmatrix} \tag{6.29}$$

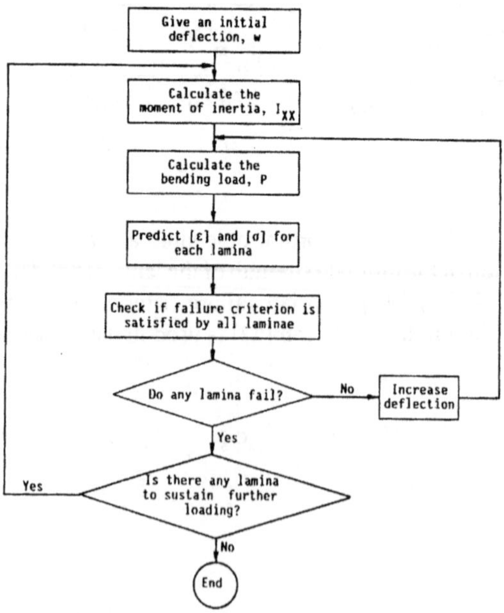

Figure 6.24. Flow chart of the design procedure for a cantilever thin-walled beam subjected to bending.

The corresponding solution for the intermediate side walls, being under combined compression/tension straining, gives

$$\begin{bmatrix} \sigma_y \\ \sigma_y \\ \tau_{xy} \end{bmatrix} = \begin{bmatrix} \pm PLY^i / I_{XX} \\ 0 \\ PD^2[1-(2Y^i/D)^2]/8I_{XX} \end{bmatrix} \tag{6.30}$$

where, D is the height of the beam cross section.

The calculation procedure can be described as in the flow chart of Figure 6.24.

6.4.5 Crashworthy Capability: Concluding Remarks

The bending process of square and rectangular tubes made of fibre-reinforced composite materials, was examined both experimentally and theoretically.

The energy absorbing mechanism during bending can be described as: microfracturing of the compressive side; buckling and crushing of the side walls; delaminated regions around the corners; simultaneous shifting of the neutral axis of the tube towards its bent tensile region.

A typical progressive failure mechanism, characterised by an initial fracture along the longitudinal axis of symmetry of the flat part of the compressive side of the tube close to the clamping edge, spreading quickly from the centre towards the corners of the tube and causing simultaneously a large number of delamination cracks at the corners just underneath the clamping device, governs the bending process. As the hinge rotation continues, the tube sustains higher loads as the corners begin to crush, whilst its compressive side and side walls begin to buckle, leading finally to an overall separation of the tube wall. The extent of tube separation is more profound for thinner specimens.

The maximum bending moment increases with increasing wall thickness. Moreover, bending seems to be greatly affected by the geometry of the tube and the process conditions. Thus, rectangular tubular components, when bent over their strong axis ($d_2/d_1 = 0.5$), show better energy absorbing efficiency. In general, rectangular tubes show better crashworthy characteristics than circular tubes of similar dimensions, mainly for large deformations; this is probably due to the deformation undergone by the corners of the tube, the increased tube strength and the improved crashworthy capacity.

The maximum bending moment seems to mainly depend upon the strength of the compressive top side and the critical strength of the tube side walls; it is well predicted analytically, see Table 6.2. The proposed analysis provides also the possibility to the designer to describe the M/θ curve in the elastic regime, see the various steps of the flow-chart in Figure 6.24. Note that the random oriented chopped strand mat composites are considered as "statistically" isotropic materials, consisting of plies with the same mechanical properties in both directions (0° and 90°). Therefore, in this case, the proposed theoretical analysis is greatly simplified.

CHAPTER 7

CIRCULAR FRUSTA

7.1 NOTATION

b_{cr} = angle of axial splits
D = outside bottom diameter of circular frustum
d = outside top diameter of circular frustum
d_c = frustum diameter at crack tip
G = fracture toughness
h = height of pulverised wedge
k = constant
L = axial length of circular frustum
L_c = height of central crack
$L_c/\cos\theta$ = length of central crack
l_s = side length of pulverised wedge
N = number of layers of circular frustum
n = number of axial splits
P = current crushing load
P_1, P_2, P_3, P_4, P_w = normal force per unit length
$\overline{P}$ = mean crushing load
P_{max} = peak load
R_{ad} = fracture energy per unit area of layers
r = radius of curvature of the frond
s = displacement, shell shortening, crush length
t = wall thickness of circular frustum
$t/\overline{D}$ = slenderness ratio
v = crush-speed
W = energy absorbed
W_T = total energy dissipated
W_s = specific energy
W_{tr} = energy required for the crush zone formation

$\alpha\ (=\alpha_1+\alpha_2)$ = angle of pulverised wedge
θ = semi-apical angle of circular frustum
μ_{s1} = static friction coefficient between frond and platen
μ_{s2} = static friction coefficient between wedge and fronds
μ_{d1} = dynamic friction coefficient between frond and platen
μ_{d2} = dynamic friction coefficient between wedge and fronds
σ_θ = tensile fracture stress
σ_o = normal stress

7.2 GENERAL

The present chapter deals with the static and dynamic axial collapse and the crush behaviour of truncated conical shells (circular frusta), made from chopped strand glass mat and polyester resin, designated as material A, [82, 83, 94, 95]. The effect of specimen geometry, i.e. thickness, diameter, length of the shell and its apical angle, on the energy absorbing capability are investigated both experimentally and analytically. Attention is directed towards the mechanics of the axial crumpling process from macroscopic point of view, for facilitating engineering design calculations of the amount of energy dissipated and the buckling loads exerted during axial quasi-static and dynamic loading. Furthermore, a study of micro-failure process during the stable collapse is also reported. Extensive microscopic investigations on the crush zone and a suitable arrangement of strain-gages on the shell wall, enable the prediction of failure mechanism characteristics, the strain distribution over the shell body and the crack length developed during crushing.

Finally, a theoretical analysis of the observed stable collapse mechanism of thin-walled circular frusta, crushed under axial static and/or dynamic loading, for calculating crushing loads and the energy absorbed during collapse, is reported. The analysis is based on experimental observations, regarding the energy absorbing collapse mechanisms developed during the crushing process. The proposed theoretical model was experimentally verified and proved to be very efficient for theoretically predicting the energy absorbing capability of the conical shells.

7.3 AXIAL COLLAPSE: STATIC AND DYNAMIC

7.3.1 Experimental

The static axial collapse was carried out between the parallel steel platens of a SMG hydraulic press at a crosshead spead of 10^{-4} m/s until a deformation of 63 mm was reached, when possible, for the specimens tested, whilst the corresponding dynamic tests were performed by direct impact on a drop-hammer with a 47 kg falling mass and maximum drop-height of 4m at velocities exceeding 1 m/s. Load/shell shortening (displacement) curves were automatically measured and recorded for both types of loading during the crushing process. The experimental set-up and the measuring devices used throughout the tests are described in detail in Chapter 5. The values of the initial peak load, P_{max} and the total energy absorbed, W_T for the axially collapsed specimens, obtained by measuring the area under the load/displacement curve, as well as the mean crushing load, $\overline{P}$, defined as the ratio of energy absorbed to the total shell shortening, and the specific energy, W_s, which is equal to the total energy absorbed per unit mass crushed (crushed volume times the density of the material), are tabulated in Table 7.1 along with the geometries of the specimens used.

Six different conical frusta forms were fabricated from laminated hard maple wooden blocks, so that 5°, 10°, 15°, 20°, 25° and 30° semi-apical angles were attained. The forms, after being machined to their final size, were coated with a layer of polyester and catalyst and cured in a standard fan-assisted furnace at a temperature of 38°C for about one hour to fill any discontinuity or surface defects. Then, the cured coated forms were properly sand-papered and polished to obtain a smooth surface and, finally, covered with a layer of release wax for facilitating the specimen removal.

In order to isolate the study of the effects of the specimen geometry from other undesirable variables, care was taken for the following criteria to be met: (i) specimens must be able to give reproducible results and (ii) they had to absorb acceptable amount of crush energy. All specimens were made by a hand lay-up technique using chopped strand glass mat preimpregnated with polyester resin. The glass used was 0.8 gr/mm² chopped fibre mat. The fibre length in the mat was 50 mm with random orientation in the plane of the mat. The resin was a blend of phthalic anhydride and maleic anhydride, esterified with propylene glycol. The catalyst used for the resin polymerization was methyl ethyl ketone peroxide. The amount of catalyst required was determined to be 1.5–3% of the resin weight. Owing to the very short time period between reduction in resin viscosity and the rapid advancement of the curing process, the specimens were formed at a low room temperature of 16°C. The resin/catalyst mixture was spread evenly on the mold and the glass mat, cut into the desired shape, was wrapped around the resin covered mold. A seam of about 20 mm wide was formed and the mixture was spread again evenly on the glass mat. For multi-layered frusta, subsequent glass layers were wrapped around the first one and rolled in, while the resin was still semi-viscous. Attention was paid to produce an even surface of the specimens by squeezing out the excess resin and getting rid of as many air bubbles as possible. Finally, all specimens were cured in a fan-assisted furnace at 38°C for one hour and then cooled slowly for about 30 min. Once removed from the mold, the shells were cut into the desired dimensions by a thin abrasive cut-off wheel and then were finished by polishing, smoothing their exterior seams and by squaring their ends. Details related to the specimen dimensions are presented in Table 7.1. The stress-strain curve of the material used, as obtained from a quasi-static tension test, is shown in Figure 5.4 (material A) of Chapter 5.

Schematic diagrams of the various failure modes recorded throughout this experimental work are indicated in Figures 7.1–7.7. A classification chart showing the areas of collapse modes and the transition boundaries from one mode to another is presented in Figure 7.8. Photographs of typical terminal collapse modes are shown in Figure 7.9.

To obtain a concept of the failure mechanism of the stable collapse mode, microscopic investigations were performed on crushed frusta after collapse to a certain amount of deformation, see Figure 7.9, using metallographic techniques. Micro-

Table 7.1: Crushing characteristics of axially loaded circular frusta.

(a) Static

Sp. No	Semi-apical angle, θ (°)	Number of layers	Thick-ness, t (mm)	Axial length, L (mm)	Outside diameter, (mm)		$t/\overline{D}$ *	$L/\overline{D}$	Crush length, s (mm)	Collapse Mode	Crushing load, P (kN)			Total energy absorbed, W_T (kJ)		Specific energy, Ws (kJ/kg)	
					Bottom end, D	Top end, d					Initial peak, P_{max}	Mean post-crushing, $\overline{P}$					
												Exper.	Theor.	Exper.	Theor.	Exper.	Theor.
1	5	2	1.6	172.7	55.6	25.4	0.040	4.26	63.5	Ia	12.2	18.4	19.5	1.172	1.238	66.7	70.5
2	5	3	2.4	152.4	55.8	29.2	0.057	3.59	63.5	Ia	29.2	27.1	24.2	1.726	1.536	63.6	56.6
3	5	4	4.3	174.2	64.8	34.5	0.052	2.12	122.4	Ia	51.3	58.6	60.3	7.180	7.380	65.8	67.6
4	5	6	5.6	179.2	73.9	42.7	0.096	3.07	76.2	Ia	77.5	86.4	89.0	6.584	6.782	64.0	65.9
5	10	1	0.8	127.0	74.7	30.0	0.015	2.43	63.5	IV	2.3	2.2	-	0.140	-	-	-
6	10	1	0.8	127.0	119.4	74.7	0.008	1.31	63.5	IV	5.7	1.2	-	0.072	-	-	-
7	10	1	0.8	144.8	175.5	124.5	0.005	0.97	63.5	IV	7.2	1.5	-	0.093	-	-	-
8	10	1	0.8	124.7	186.2	142.2	0.005	0.76	63.5	IV	6.4	1.6	-	0.098	-	-	-
9	10	1	0.8	279.4	284.2	185.7	0.003	1.19	20.2	IV	3.6	-	-	0.042	-	-	-
10	10	2	1.7	152.4	79.2	27.0	0.030	2.87	63.5	Ia	17.9	18.3	19.2	1.169	1.219	47.5	49.5
11	10	2	1.7	149.9	132.1	79.2	0.016	1.42	39.0	III	42.8	2.9	-	0.113	-	-	-
12	10	2	1.7	150.5	132.3	79.2	0.016	1.42	63.5	Ic	47.3	12.5		0.822		17.0	
13	10	2	1.7	149.9	129.0	76.2	0.015	1.46	63.5	Ic	37.0	14.2		0.926		21.6	
14	10	3	2.3	159.9	81.3	27.9	0.040	2.93	63.5	Ia	26.0	31.2	28.7	1.986	1.822	59.8	54.7
15	10	4	4.3	164.1	92.4	34.8	0.068	2.57	117.2	Ia	42.7	59.0	63.5	6.920	7.442	57.1	61.4
16	10	6	7.4	164.6	89.2	31.2	0.123	2.73	115.9	Ia	124.9	124.0	129.7	14.380	15.032	66.5	69.5
17	15	2	1.7	152.4	111.8	30.2	0.024	2.14	63.5	Ia	14.5	17.9	20.3	1.137	1.289	34.7	39.3
18	15	2	1.7	152.4	193.5	111.8	0.011	1.00	63.5	III	31.1	6.0	-	0.382	-	-	-
19	15	2	1.6	142.2	187.9	111.8	0.011	0.95	63.5	III	49.1	4.9	-	0.308	-	-	-
20	15	3	2.9	142.2	105.4	29.2	0.038	2.11	63.5	Ia	29.4	33.3	29.8	2.121	1.892	46.4	41.4
21	15	4	5.1	159.0	126.0	40.6	0.061	1.91	74.6	Ia	49.0	66.4	69.0	4.950	5.147	50.9	52.9
22	15	6	6.6	151.9	109.2	27.9	0.096	2.16	106.8	Ia	99.2	125.7	130.6	13.430	13.948	65.1	67.6
23	20	2	1.5	142.9	132.1	27.9	0.019	1.77	63.5	Ib	7.3	8.2	10.2	0.522	0.647	15.3	19.0
24	20	3	2.3	139.7	134.6	33.0	0.027	1.67	38.2	IIb	20.6	31.1	-	1.187	-	-	-
25	25	2	1.3	142.2	164.3	31.7	0.013	1.45	63.5	IIa	4.5	4.7	-	0.296	-	-	-
26	25	3	2.3	134.6	160.0	34.3	0.023	1.39	19.1	IIb	31.5	35.6	-	0.680	-	-	-
27	30	2	1.7	139.7	189.7	28.6	0.015	1.28	63.5	IIa	4.9	6.7	-	0.424	-	-	-
28	30	3	2.3	134.6	193.8	38.5	0.020	1.16	19.0	IIb	19.1	9.0	-	0.171	-	-	-

Table 7.1 (cont.)

(b) Dynamic

Sp. No	Semi-apical angle, θ (°)	Number of layers	Thick-ness, t (mm)	Axial length, L (mm)	Outside diameter, (mm)		$t/\overline{D}$ *	$L/\overline{D}$	Crush speed, v (m/s)	Crush length, s (mm)	Collapse Mode	Crushing load, P (kN)			Total energy absorbed, W_T (kJ)		Specific energy, Ws (kJ/kg)	
					Bottom end, D	Top end, d						Initial peak, P_{max}	Mean post-crushing, $\overline{P}$					
													Exper.	Theor.	Exper.	Theor.	Exper.	Theor.
29	5	2	2.2	163.7	58.9	31.8	0.049	3.61	7.0	73.3	Ia	10.8	13.6	14.3	0.997	1.048	33.1	34.8
30	5	3	3.1	167.2	61.7	32.3	0.066	3.56	7.0	41.2	Ia	38.9	26.2	24.0	1.080	0.989	44.6	40.8
31	5	4	4.3	167.0	63.6	36.5	0.086	3.34	8.1	42.6	Ia	82.8	35.6	36.8	1.519	1.567	42.0	43.3
32	5	5	5.2	168.0	67.0	39.6	0.098	3.15	8.1	31.9	Ia	67.3	43.2	40.0	1.377	1.276	39.6	36.7
33	5	6	5.7	167.2	66.5	38.3	0.109	3.03	8.1	26.0	Ia	104.1	50.8	52.9	1.320	1.375	44.2	46.0
34	10	2	2.1	159.0	83.8	32.9	0.036	2.72	7.0	66.3	Ia	11.1	15.9	17.5	1.052	1.160	30.8	34.0
35	10	3	2.9	155.5	86.8	33.0	0.048	2.60	7.0	40.3	Ia	22.1	25.4	21.0	1.023	0.846	35.3	29.2
36	10	4	4.7	167.9	96.5	41.2	0.068	2.44	8.1	28.7	Ia	91.0	47.8	50.6	1.341	1.452	35.2	38.1
37	10	5	5.1	167.3	92.1	36.1	0.080	2.61	8.1	25.5	Ia	76.8	48.7	51.1	1.239	1.303	37.1	39.0
38	10	6	6.5	158.7	91.7	38.0	0.100	2.44	8.1	22.5	Ia	125.0	65.0	69.2	1.461	1.557	39.5	42.1
39	15	2	2.1	137.5	109.4	34.6	0.029	1.91	7.0	75.2	Ib	14.7	13.6	15.3	1.035	1.151	20.9	23.2
40	15	3	2.9	164.1	120.0	31.2	0.038	2.17	7.0	44.1	Ib	35.3	23.6	20.1	1.040	0.886	25.2	21.5
41	15	4	4.7	152.0	118.1	34.5	0.062	1.99	8.1	47.5	Ia	47.1	30.6	34.3	1.453	1.629	25.8	28.9
42	15	5	5.2	160.5	119.8	31.9	0.069	2.12	8.1	31.4	Ib	31.3	44.1	43.2	1.384	1.356	27.0	26.5
43	15	6	6.8	154.5	124.3	40.0	0.083	1.88	8.1	22.6	Ib	87.4	70.0	73.7	1.401	1.666	27.3	32.5

* $\overline{D}=(D+d)/2$

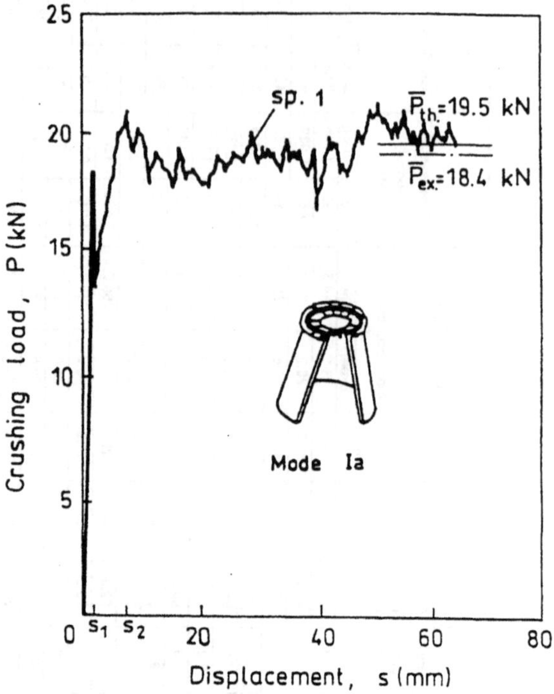

Figure 7.1. Schematic diagram of collapse Mode Ia and load/deflection characteristics of specimen 1 (see Table 7.1).

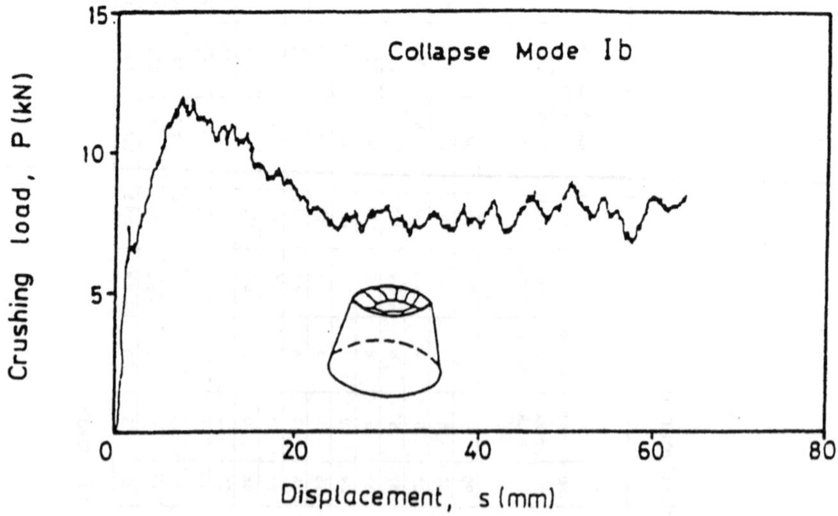

Figure 7.2. Schematic diagram of collapse Mode Ib and load/deflection characteristics of specimen 23 (see Table 7.1).

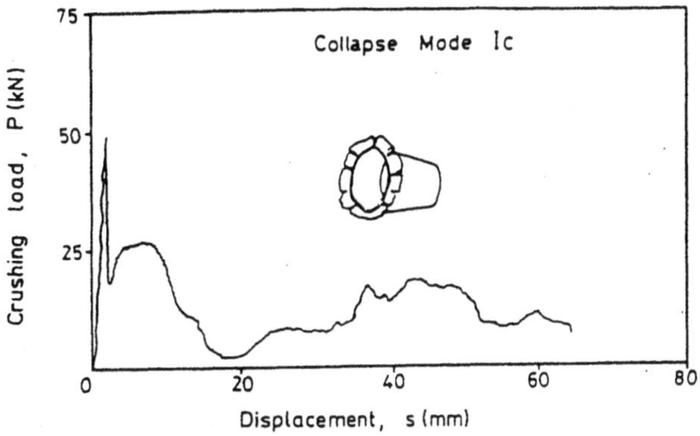

Figure 7.3. Schematic diagram of collapse Mode Ic and load deflection characteristics of specimen 12 (see Table 7.1).

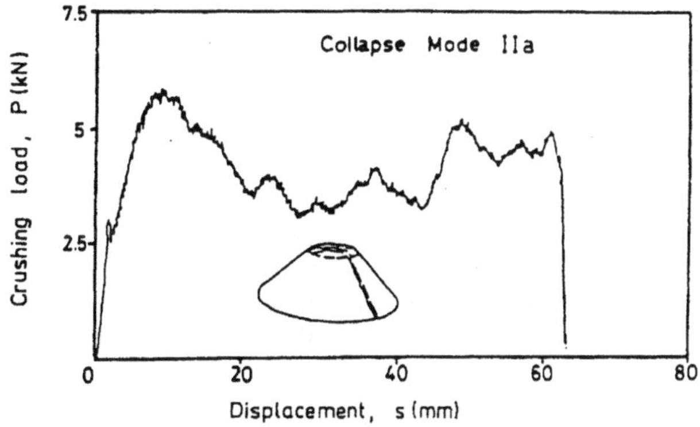

Figure 7.4. Schematic diagram of collapse Mode IIa and load/deflection characteristics of specimen 27 (see Table 7.1).

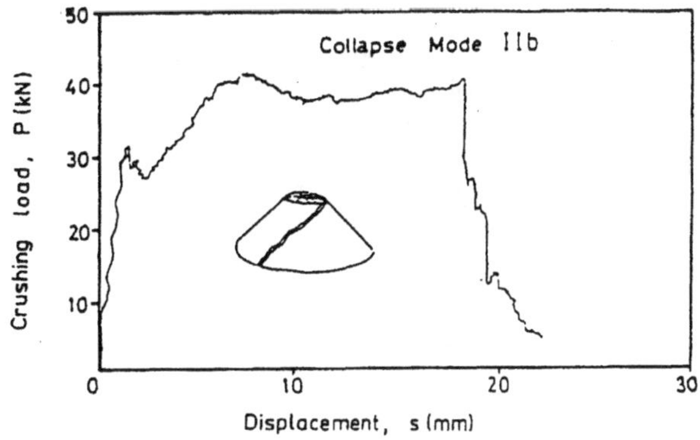

Figure 7.5. Schematic diagram of collapse Mode IIb and load/deflection characteristics of specimen 26 (see Table 7.1).

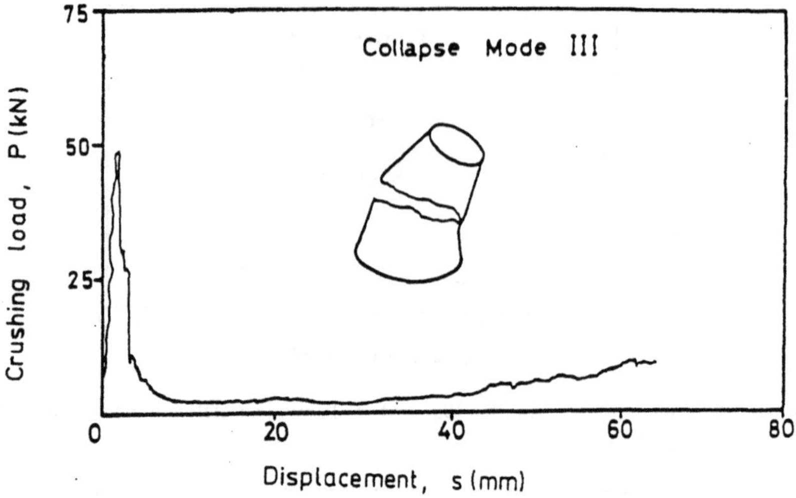

Figure 7.6. Schematic diagram of collapse Mode III and load/deflection characteristics of specimen 19 (see Table 7.1).

graphs of the crush zone were obtained by examining metallographic specimens, cut-off from the damaged region of the compressed frusta, on a Unimet metallographic optical microscope equipped with photographic facilities. To prepare the metallographic specimens, a longitudinal strip at the damaged region was removed from the shell wall using a small fret-saw with a sharp, fine blade, and encapsulated

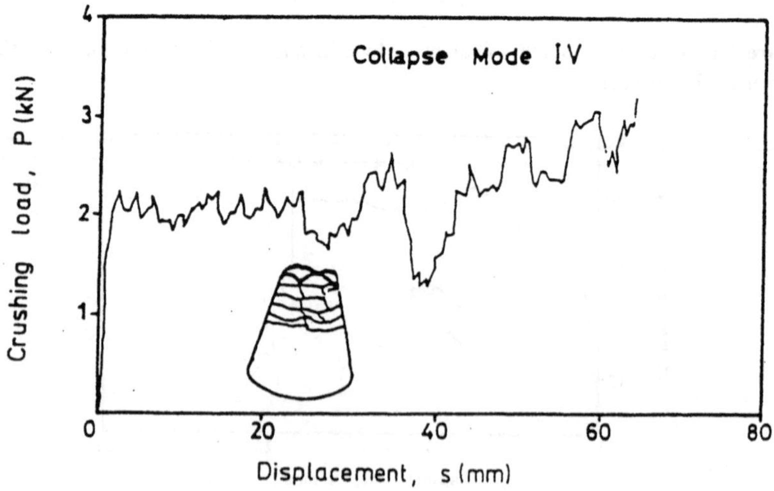

Figure 7.7. Schematic diagram of collapse Mode IV and load/deflection characteristics of specimen 5 (see Table 7.1).

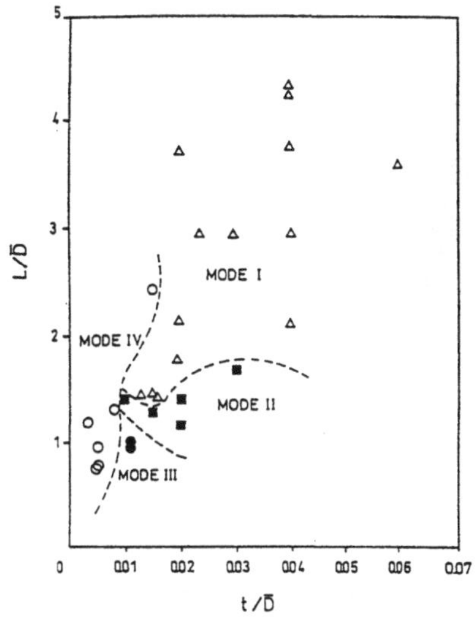

Figure 7.8. Classification chart showing the areas of collapse modes and transition boundaries from one mode to another for composite circular frustra.

in potting resin in conventional metallographic plastic mould. The relevant surface was then prepared successively on 200, 400, 600 and 1200 grit abrasive wheels and polished using quarter micron alumina paste. Increased visual contrast between fibres and resin was attained by orienting the specimen such that the direction of polishing was perpendicular to the shell wall during the stage of finishing. The polished

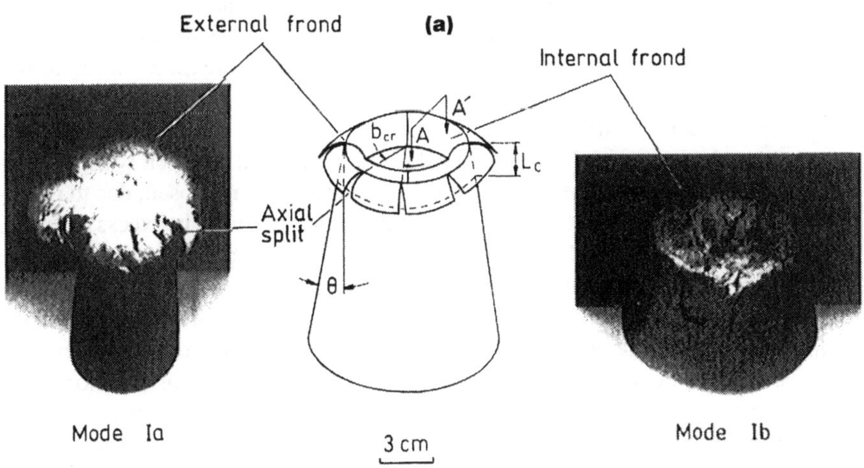

Figure 7.9. (a) Macroscopic side view of the collapse modes Ia and Ib, and the configuration of failure mechanism of Mode Ia.

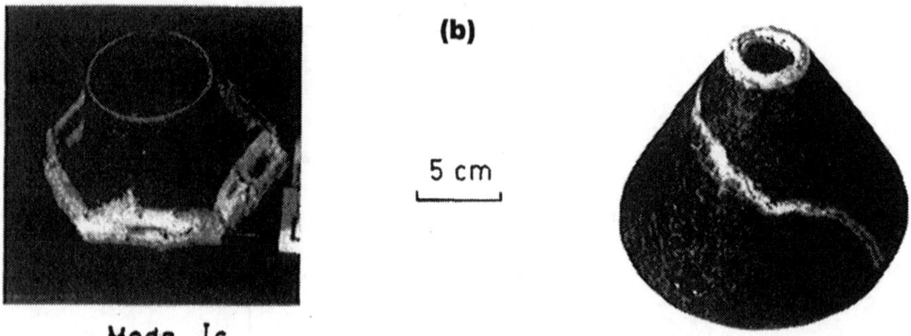

Figure 7.9 (continued). (b) Macroscopic side view of the collapse modes Ic and IIb.

specimen was then etched by immersing it for 10 seconds in a reagent consisting of 50% solution of hydrofluoric acid, thoroughly washed in distilled water and afterwards for 30 minutes in a solution of sodium bicarbonate in water to ensure that no trace of hydrofluoric acid was remained. The polished and etched cross-sections were finally photographed at magnifications higher that 50 on the optical microscope. Relevant micrographs of the specimens tested are presented in Figure 7.10, see also the details in Table 7.1.

An attempt was made to obtain information about crack propagation along the wall of the shell by monitoring the strain fields in the shell as the crush zone progresses during collapse (Mode Ia); for this purpose, fifteen rosette strain gages were located at various positions on the inner surface of a 10° frustum, as shown in Figure 7.11. Electro-resistance strain-gages (90 degree "tee" rosettes and 45 degree rectangular rosettes) of 3% maximum strain and with 350 Ohm resistance were chosen to accommodate the reduced heat conductivity of polymer based composites and the high excitation voltage used during data-acquisition. Selection of the 90 degrees "tee" rosettes was made, based on the assumption, that the composite material tested was transversely isotropic and the principal stress directions were along the generator and the hoop directions; however, to check this assumption one 45 degrees rectangular rosette was also mounted. For mounting the strain-gages on the shell wall, the general purpose laboratory adhesive methyl-2-cyanoacrylate was chosen, due to its curing temperature and elongation capability (high-elongation tests in excess of 30.000 microstrain can be run, whilst curing is obtained at room temperature). To overcome difficulties, usually encountered during the procedure of mounting strain-gages on composite materials, short wires were soldered onto the strain-gages using an aluminium block heat sink, whilst vacuum pressure was applied to hold the strain- gage onto the block. A neutralizer was then used to clean the specimen surface thoroughly and the strain-gages were mounted at the locations selected. The lead wires were connected to the gage based on three wire hook up. All strain gages were checked for the presence of air bubbles between the bonding surface and it was found to be within tolerable limits; they were calibrated for maximum resolution after being connected to a strain-gage amplifier unit. During the collapse process, the strain readings obtained

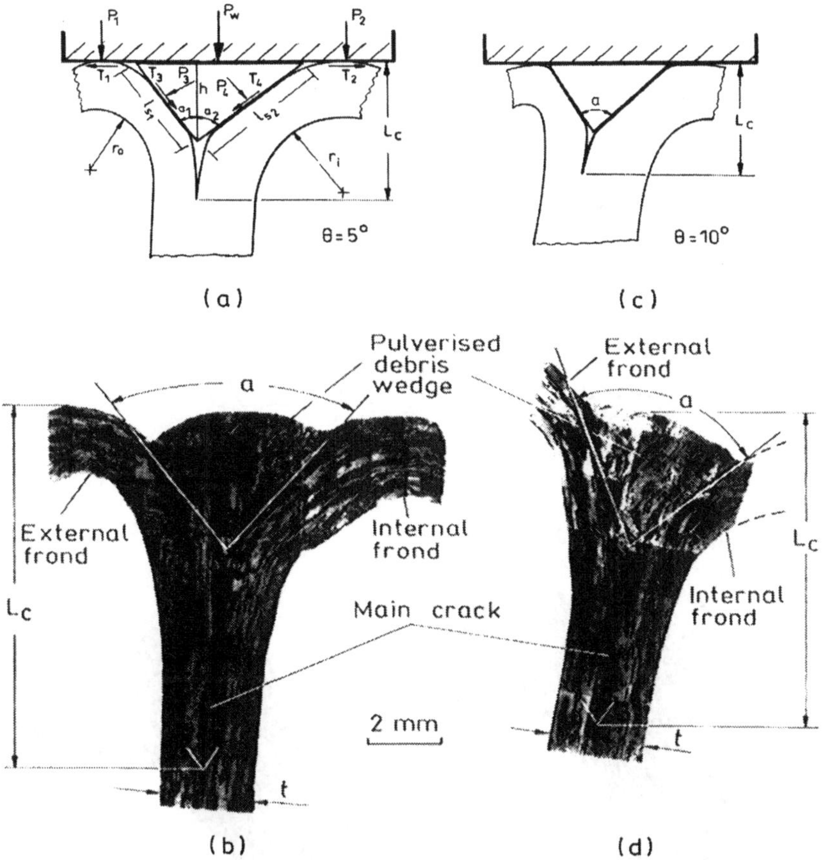

Figure 7.10. (a) Configuration of the crush zone (Mode Ia) through the wall thickness of a 5° specimen (cross section AA′ in Figure 7.9), (b) micrograph showing microfailures in the crush zone (Mode Ia) for a statically loaded 5° frustum (sp. 2; see Table 7.1), (c) configuration of the crush-zone (Mode Ia) through the wall thickness of a 10° specimen (cross section AA′ in Figure 7.9), (d) micrograph showing microfailures in the crush-zone (Mode Ia) for a statically loaded 10° frustum (sp. 14; see Table 7.1).

were evaluated using the KEITHLEY data-acquisition system and, therefore, the related strain distributions were determined; plots obtained from this series of measurements are shown in Figures 7.12–7.17.

The same procedure was followed for deeper understanding of the crushing behaviour of thick circular frusta with a semi-apical angle greater than 20°, which failed by global splitting after a short distance of crushing (Mode II of collapse), absorbing, in this manner, small amounts of energy. Strain-gages were used to gain information on the state of stress at certain locations, mainly at the bottom and middle of the tube wall. The bottom gages were placed in such a manner to give insight into the mecha-

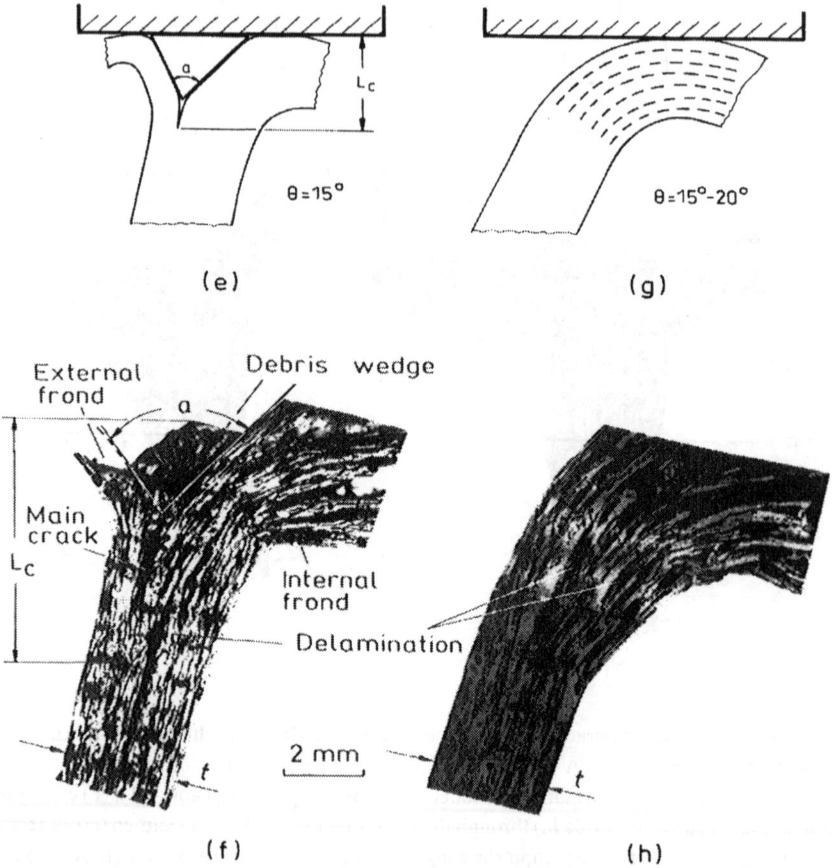

Figure 7.10 (continued). (e) Configuration of the crush-zone (Mode Ia) through the wall thickness of a 15° specimen (cross section AA′ in Figure 7.9), (f) micrograph showing microfailures in the crush-zone (Mode Ia) for a statically loaded 15° frustum (sp. 20; see Table 7.1), (g) configuration of the crush-zone (Mode Ib) through the wall thickness of a 15° specimen (cross section AA′ in Figure 7.9), (h) micrograph showing microfailures in the crush-zone (Mode Ib) for a dynamically loaded 15° frustum (sp. 40; see Table 7.1).

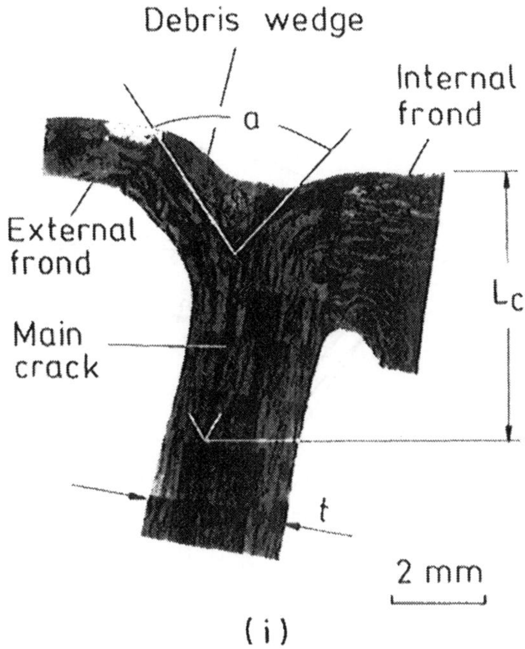

Figure 7.10 (continued). (i) Micrograph showing microfailures in the crush-zone (Mode Ia) for a dynamically loaded 10° frustum (sp. 35; see Table 7.1).

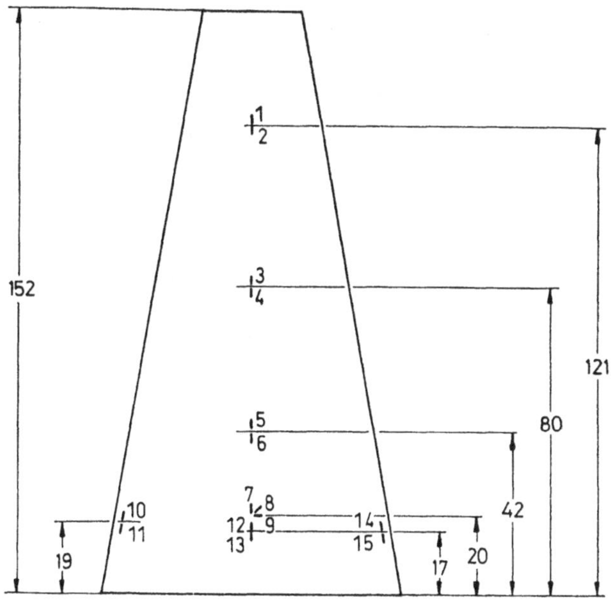

Figure 7.11. Arrangement of strain-gages on a 10° frustum.

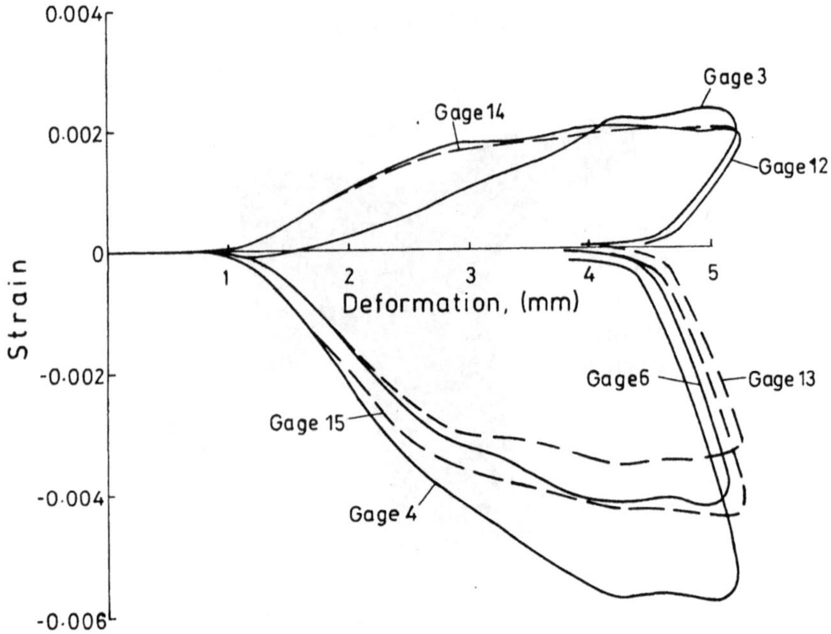

Figure 7.12. Strain distributions during stage I of loading.

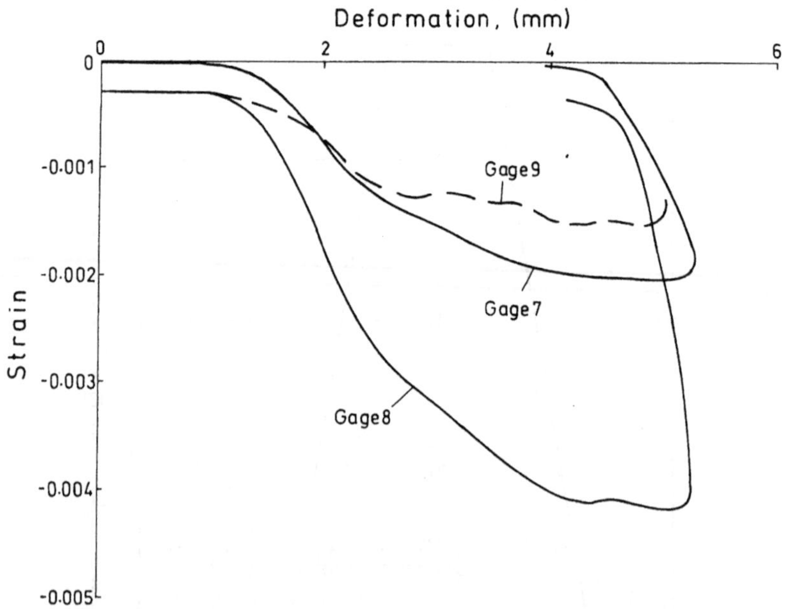

Figure 7.13. Strain distributions during stage I of loading showing direction of principal stresses during axial compression.

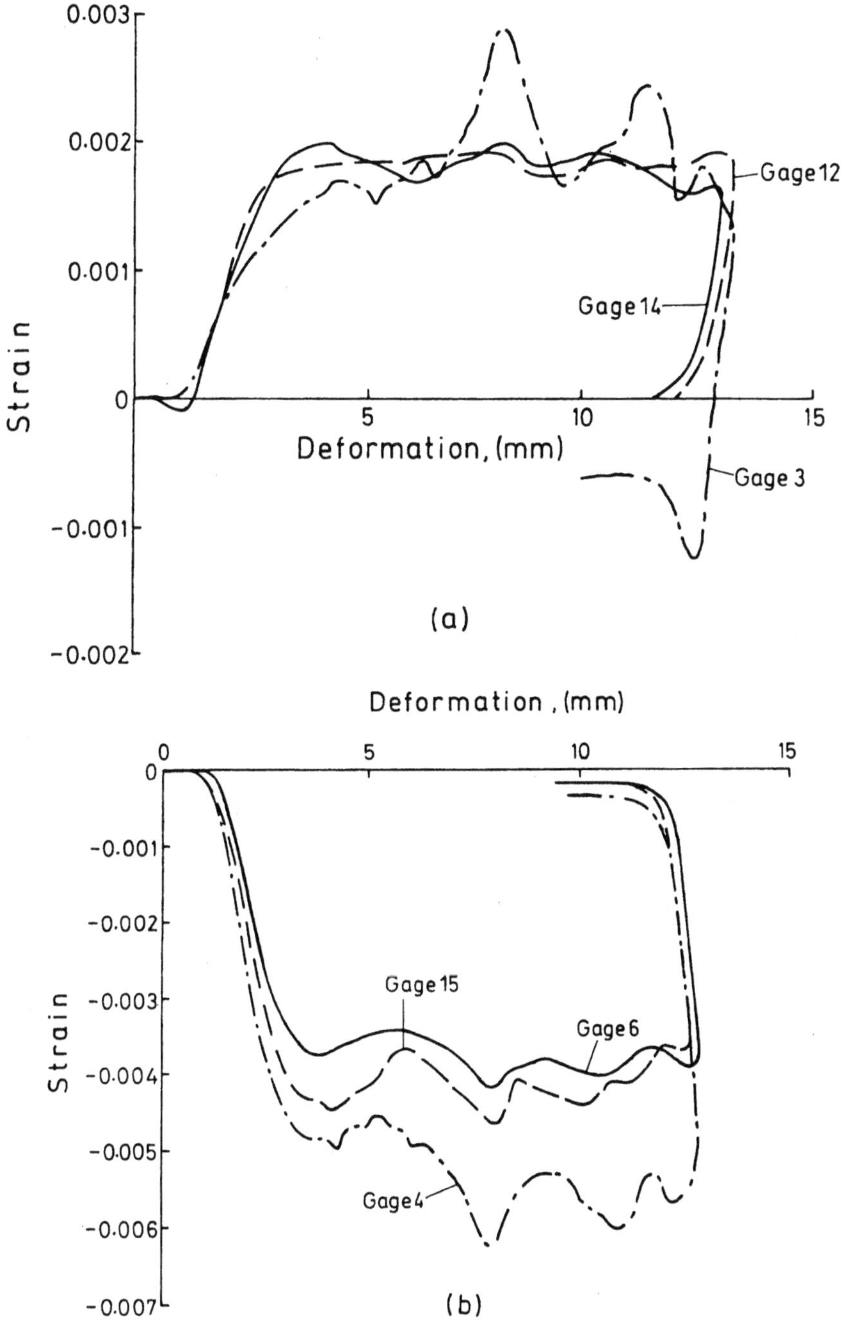

Figure 7.14. (a) Strain distributions during stage II of loading in the hoop direction, (b) Strain distributions during stage II of loading in the longitudinal direction.

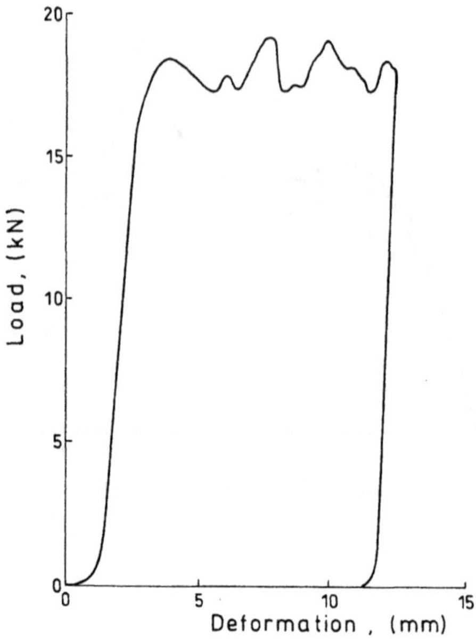

Figure 7.15. Load/deflection curve for the specimen shown in Figure 7.11 as obtained during the loading stage II.

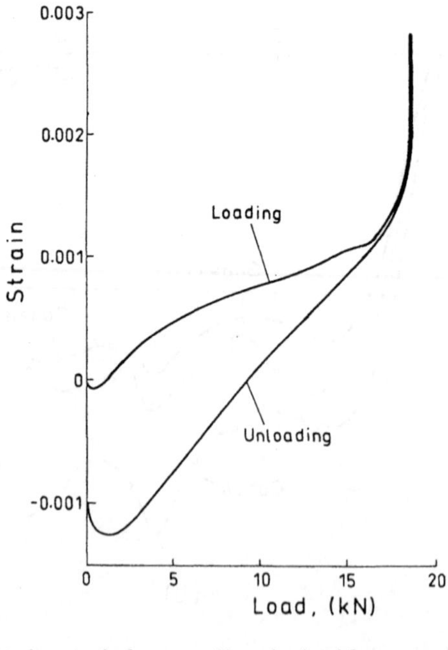

Figure 7.16. Strain/load curve during stage II as obtained from gage 3 showing creep development.

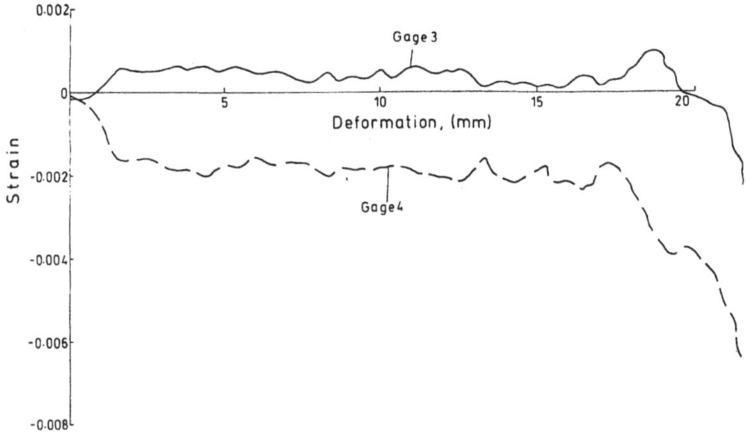

Figure 7.17. Strain distributions obtained by gages 3 and 4 during loading stage III.

nism of the splitting initiation at this area and to simultaneously bound the critical area, whilst the middle gages were used to provide information on the state of stress close to the main crush zone. The strain gaged specimens were 25° four-layered frusta, similar to specimens 25 and 26 in Table 1.

The strain gaging was performed in two stages and the gage placement for each stage is shown in Figures 7.18 and 7.19, respectively. In the first stage, biaxial rosettes of 120 Ω resistance were placed only on the inside surface of the frustum, the even numbered gages lie axially (longitudinal direction) and the odd numbered ones transverse to it (circumferential direction), see Figure 7.18. In the second stage, biaxial rosettes of 350 Ω resistance were placed opposite to each other on the inside and outside surfaces of the specimen, see Figure 7.19, so that bending effects near the bottom edge could be analysed through its wall thickness.

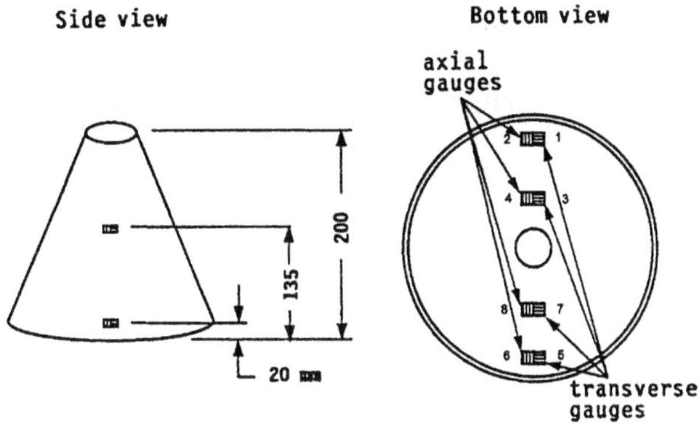

Figure 7.18. Arrangement of strain-gages on a 20° frustum (stage I).

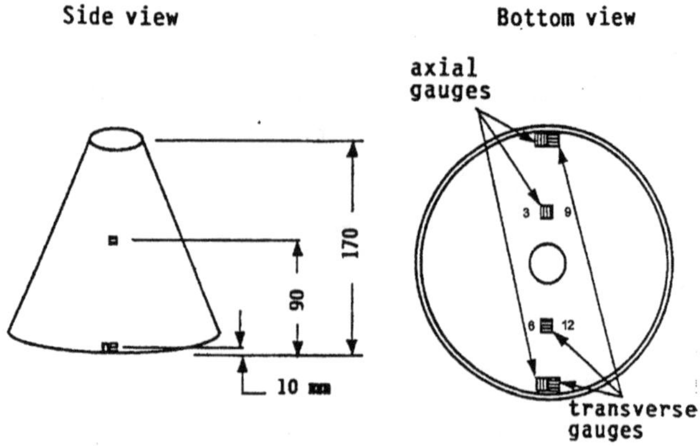

Figure 7.19. Arrangement of strain-gages on a 20° frustum (stage II).

7.3.2 Failure Mechanisms: Experimental Observations

COLLAPSE MODES AT MACROSCOPIC SCALE

Conical frusta made from composite materials and subjected to static axial collapse were found to collapse in modes considerably different than those observed in metallic and thermoplastic structures, see References [8, 106–108]. The brittle nature of both fibres and resin ensures that composite materials do not undergo the characteristic for ductile metals and PVC plastic deformation. On the contrary, the mechanism of fracture and fragmentation dominates rather the crushing phenomenon. In general, the failure modes observed throughout the tests are greatly affected by the shell geometry, the arrangement of fibres, the properties of the matrix and fibres of the composite material and the stacking sequences.

Four distinct modes of collapse were observed, identified and classified as follows; see Table 7.1(a) and also References [82, 83]

(a) End-Crushing Mode (Mode I)

This type of failure is mainly characterised by stable collapse of the specimen and the formation of continuous fronds, which spread radially into three different directions, defining the following modes of failure:

- Mode Ia: Collapse initiates at the narrow end of the conical shell and, as delamination progresses, the inner layers of the shell inverse inwards and the outer ones outwards by progressive crushing in the form of "mushrooming" failure; see Figures 7.1 and 7.9(a).
- Mode Ib: Collapse starts also at the narrow end of the shell and the shell wall inverses inwards, see Figures 7.2 and 7.9(a). This failure mode is observed for specimens of higher semi-apical angles.

- Mode Ic: Progressive inversion of the shell wall outwards, starting at the large end of the frustum, was observed for some specimens of high semi-apical angles (θ=10°); see Figures 7.3 and 7.9(b).

The main observations regarding the collapse Modes Ia–Ic may be listed as follows:

- The externally formed fronds inverse freely outwardly with the simultaneous development of a number of axial splits followed by splaying of the material strips. The length of axial splits probably defines the effective column length of the material strips undergoing loading.
- The internally formed fronds are turned inside the tube wall; firmly compacted hoops of material, composed by transversely arranged fibres, are developed, which are tightly packed and further compacted, constraining, in this manner, the frond to fold inwards. No axial tears are apparent in the internal fronds, which are more continuous than their external counterparts.

(b) Transition Mode of Failure (Mode II)

In this kind of failure, collapse initiates at the top end and, as delamination progresses, the same type of failure described in (a) above follows, but at a certain point of deformation, the extent of progressive crushing is limited by shell failure, associated with the formation of rapidly propagating cracks to some distance ahead of the frustum end, leading to unstable collapse modes, defined as:

- Mode IIa: A longitudinal crack is formed as shown in Figure 7.4(a)
- Mode IIb: A 45° spiral crack, is formed, propagating along the shell circumference; see Figures 7.5 and 7.9(b).

(c) Mid-Length Collapse Mode (Mode III)

This mode shows the characteristic features of brittle fracture, see Figure 7.6. Fracture starts at a distance from the frustum end, approximately equal to the mid-height of the frustum, and involves catastrophic failure by cracking and separation of the shell into irregular shapes, probably due to local severe shear straining of the tube wall. Note that this failure mode shows characteristics similar to the Euler column-buckling of very thin metallic and PVC tubes [8].

(d) Progressive Folding Mode (Mode IV)

This kind of collapse is by progressive folding with the formation of sharp hinges, similar to the collapse of metallic and PVC tubes, see References [8, 106–108]. A series of folds, or fracture hinges, develop as the conical shell is crushed, see Figure 7.7.

The overall collapse mode of conical frusta, fabricated from chopped strand mat laminate (composite material A), mainly depends upon the shell wall thickness (number of layers), its semi-apical angle and its mean diameter, $\overline{D}$. Two-layered specimens, with fairly small semi-apical angles 5°–15°, showed a more profound tendency to collapse in a stable mode of failure, i.e. Modes Ia and Ib, than the 25° and 30° frusta, which preferably follow the Mode IIa of collapse; compare the related specimens in Table 7.1. An increase in shell thickness, i.e. in the case of three-layered specimens, resulted in the same stable mode of collapse (Mode Ia), when collapsing 5°, 10° and 15° frusta, whereas the 20°, 25° and 30° frusta failed in a mixed mode of collapse (Mode IIb). From these observations, a transition region between stable and unstable mode of collapse, due to the effect of the semi-apical angle, may be designated. The transition semi-apical angle of the frustum, related to this transition region, was identified to fall between 15° and 20°.

Frusta, with a wall thickness related to one-layered shell, collapsed by progressive folding (Mode IV) and, therefore, a critical transition shell thickness may be defined. With increasing wall thickness, the collapse mode for the 20° and 25° frusta changed from Mode I to Mode IIb, while the 30° frusta changed from Mode IIa to Mode IIb. However, the collapse modes remained the same for the 5°, 10° and 15° frusta, even with a thickness increase.

The effect of the mean diameter, $\overline{D}$ of the shell on the mode of failure is rather negligible. Conical frusta with large mean diameters were more prone to the catastrophic Mode III type of failure, see Table 7.1(a).

The characteristic modes of collapse, which are observed throughout dynamic axial tests, can be identified and classified as stable and unstable collapse modes. It must be noted that the failure modes encountered during the present impact testing are only stable modes and similar to the observed during the corresponding quasi-static ones, see Table 7.1(b).

The experimentally obtained deformation modes of all specimens tested are classified in respect to the ratios $t/\overline{D}$ and $L/\overline{D}$ and are presented in Figure 7.8. Distinct regions, characterising the various deformation modes occurred, and the transition boundaries from one mode to another are presented, providing useful information about the collapse mode of the conical frusta used an energy absorbers.

COLLAPSE MODES AT MICROSCOPIC SCALE

As stated above, progressive collapse, designated as Mode I macroscopic type of deformation, see Figures 7.1–7.3 and 7.9, offers the most effective and efficient mode of energy absorption; therefore, as far as the microfailure mechanisms, encountered for the plastic collapse, i.e. crack initiation and propagation as collapse progresses, is concerned, attention is mainly focused on the major features of the crush zone of this microscopic failure mode. An attempt was also made to relate the levels of energy absorption to the microstructural failure process. Note that progressive crushing is characterised by the formation of a well defined crush zone, which travels along the frustum wall longwise. The main features of this crush zone are, see Figure 7.10(b) and References [82, 94, 109].

- An annular wedge of highly fragmented material, forced down axially through the shell wall

- an intrawall microcrack, which develops ahead of the crush zone at the apex (tip) of the annular wedge and propagates at a rate approximating the compression rate
- two continuous fronds (internal and/or external) as a result of the plies delamination in the crush zone, mainly caused by the central bundle wedge, which spreads radially inwards and outwards the wall of the frustum
- a severely strained zone (compressive-tensile zone), which extends between the central crack and the shell wall edges, showing a combined tensile-compressive type of deformation

The crushing process of fibre-reinforced composite shells is a cyclic process characterised by the development of a wedge-shaped region, accompanied by an interlaminar crack propagating between plies in the crushed region and forming two lamina bundles; they resist the applied load and buckle when the applied load or the length of the lamina bundle reaches a critical value. The crushing pattern depends upon the maximum strain at failure of the fibre and the matrix, as well as on the ply orientation and the stacking sequence of the material. To identify the fracture process the following secondary mechanisms may be noted, see also Figure 7.10:

- Bending of the axial laminae
- Splitting of the outer axial lamina and tensile fracturing of the outer hoop lamina, due to increasing circumference of the frustum
- Crushing and multiple fracturing of inner laminae, due to reducing shell circumference and, therefore, volume availability
- Cracking down of the central regime of the frustum wall, due to compressive loading
- Fracturing between hoop and axial laminae, as the latter expands outwards/inwards and downwards
- Accumulation of debris in the central region of the wall of the frustum

Delamination occurs as a result of shear and tensile separation between plies, see Figure 7.10. Note that each ply can withstand its own bending load and, additionally, the loads exerted by the plies beneath it. As bending occurs, the axial laminae split into progressively thinner layers, forming, therefore, translaminar cracks normal to the fibre direction mainly due to fibre buckling, finally resulting either in fibre fracture or in intralaminar shear cracking, which splits the laminae into many thin layers without fibre fracture.

The intrawall crack, see Figure 7.10, starts in the frustum wall at a certain distance from the top of the frustum/platen interface and it is of predominantly tensile nature. The mechanism, which governs its formation, is probably related to the lateral expansion of the shell wall during the axial compression, resulting in a number of fissures, which are developed at the platen/crush zone interface; the latter, facilitated thereafter by the piercing of the debris wedge formed, develops as a steady state travelling intrawall crack. Resistance to the crack propagation along the central region of the wall is offered by the radial compressive stiffness of intact internal material and by the tensile hoop strength of the outer plies and the interply bonding. Further crushing results in compression of the axial laminae into the cavity, comminuting, therefore, the material into small particles of debris to such an extent that the original fracture surfaces are completely destroyed; a complex process of fibre bundles bending, interpenetrating and sliding over each other, because of the limited volume in the shell cavity, probably contributes to such a debris formation.

As a result of the above-mentioned crushing mechanism, a triangular shaped wedge, consisting of fine debris, is formed, situated at the open end of the intrawall crack at the platen/crush zone interface, see Figure 7.10; it comprises of small pieces of broken glass fibres amongst small particles of powdered resin and composite. Since the diameter of the wedge increases, as the cone crushing progresses, it is evident that the size of this wedge increases during crushing.

The externally displayed frond, see Figures 7.10(a)–(f), was divided into a number of curled strips with varying width and it depends mainly on shell geometry and stacking conditions. The axial splitting observed can be attributed to the radial expansion of the compressed frustum, whilst expansion in the hoop direction of the shell is negligible, because of the detrimental hoop strength of the composite material. White horizontal lines, spaced at about 5 mm intervals, are visible on the concave surface of the strips. The short length of the strips indicates that most of the frond did not remain fixed on the shell wall during crushing, whilst in the rest of it delamination occurred between the layers of the cloth of the composite material. As the outer axials spread outwards, the hoops must expand subjected to tensile loading, resulting, therefore, in cracks grown parallel to the fibres, close to the fractured region. Some hoop fibre pull-out is visible along the edges of the strip, whilst in some cases, fibres were transversely debonded from the resin matrix and a patch of complementary resin furrows was labelled. Interlaminar cracks, observed between every layer, are a consequence of the splaying of the strips of material, as they curve outwards from the frustum.

The internally formed frond, see Figure 7.10, was compacted inside the frustum in a rather irregular shape, exceeding a considerable amount of crushing and fracturing, together with a degree of whitening. It is composed of long, thin, irregular, curved strips of material. The circumferential convolutions of the material indicate the occurrence of compressive failure of the hoop fibres. Interlaminar splitting is also observed and retains a degree of through-thickness integrity. Cracks were formed on their upper surfaces due to tension, whilst the lower sides underwent repeated cracking by compression.

After the completion of the central crack and the formation of the annular wedge, and during the development of the two fronds, the combined action of the compressive upwards force and the tensile and compressive forces, due to the bending moment in a direction outwards from the crack axis, causes the material to fail in several ways in the nearby regions of the frustum, depending on the shell geometry and the stacking conditions of the composite material. Thus, the mechanisms observed can be designated as, tensile failure through the crush zone, compressive shear failure or shear failure following tensile straining.

The crack tip propagates preferably through the weakest regions of the structure of the composite material, i.e. through resin-rich regions or boundaries between hoop fibres, resulting in their debonding or through the interface between hoop and axial plies causing delamination.

Based on the experimental observations, the effect of shell geometry on the shape of the crush zone is shown schematically in Figures 7.10(a), (c) and (e). With increasing semi-apical angle of the frustum, the position of the intrawall crack moves towards the outside edge of the shell wall, increasing in this manner the thickness of the inner frond and simultaneously resulting in a positioning of the annular wedge, mainly above it. The dimensions of the triangular cross section of the pulverised wedge, as well as the crack length, are greatly reduced for higher semi-apical angles and, therefore, during crushing an internal and an external frond are formed without bundle wedge formation, whilst a shear zone is developed in both sides of the central crack. In the case of the Mode Ib, progressive collapse is created by successive shearing of the region near the narrow end of the shell. From measurements of the statically loaded specimens, the wedge angle, α was estimated to about 80°, 70° and 60° for 5°, 10° and 15° frusta, respectively.

The effect of the semi-apical angle of the frustum on the energy absorption of the collapsing shell can be related to the microstructural failure. From Figure 7.10(a) it is obvious that the radius of curvature of the internal frond, r_i increases with increasing semi-apical angle, whereas the external frond is constrained to deform through a tighter radius, r_o. A great amount of the absorbed energy for this mode of crushing is due to the deformation and fracture of the internal and external fronds, as they are bent over. An increase of the radius of curvature results in a decrease of the energy absorbed, due to this micromechanism of deformation and failure; consequently, as the semi-apical angle of the frustum increases the energy absorbed by frond deformation decreases.

In general, the microfracture mechanism of Mode Ia of collapse is similar for statically and dynamically loaded shells. The only differences encountered are associated with the shape of the pulverised wedge and the microcracking development.

In the case of the dynamically loaded circular frusta, the main cracking caused by the wedge is smaller in size, whilst the position of the wedge, which is smaller in size, and of the tip of the crack are shifted towards the outside wall surface of the impacted shells, as compared to statically loaded ones; compare Figures 7.10(f) and 7.10(i). Therefore, transition from Mode Ia to Mode Ib occurs for smaller semi-apical angles for dynamical loaded frusta, as compared to the related static ones, see Table 7.1. From measurements of the collapsed impacted shells, the wedge angle was estimated to about 75°, 60° and 50° for 5°, 10° and 15° frusta, respectively.

STRAIN DISTRIBUTION AND DETERMINATION OF CRACK LENGTH

(a) Progressive Collapse Mode (Mode Ia)

In order to determine the mechanical response of the conical shell during the crushing process, the strain gaged specimen, see Figure 7.11, was subjected to a three-stage compressive loading:

- Stage 1: The shell was loaded up to the point when fragmentation began; a corresponding shortening of 5.5 mm of the shell was obtained; see Figures 7.12

and 7.13. In this manner, the shell behaviour in the pure elastic regime can be studied

- Stage 2: After unloading, the specimen was re-loaded and crushed until both the linear regime and a highly serrated post-buckling load-deflection curve was developed, corresponding to an additional axial shortening of 12 mm, see Figures 7.14–7.16, enabling, therefore, the investigations to exceed beyond the elastic zone. At the end of this loading stage, the specimen was again unloaded.
- Stage 3: In this stage, the strain field was monitored very closely, as the crush zone approached the strain-gages. Thus, loading was done in incremental steps with advance increments of 5 mm. Load increments were continued until the gages 3 and 4 were permanently damaged, i.e. up to a total compression of 22 mm, see Figure 7.17. Note that the strain-gages 5,10 and 11 failed to give strain measurements.

As expected, all gages measuring strains in the longitudinal direction (gages 2, 4, 6, 8, 11, 13 and 15) showed compressive strains, whilst the corresponding strain-gages measuring hoop strain (gages 1, 3, 5, 10, 12 and 14) recorded tensile strains; see Figures 7.12–7.14.

During the loading stage in the elastic regime (stage 1), it was determined that away from the crush zone the material, besides retaining its original integrity, behaves also elastically. Strains returned to zero after unloading, see Figures 7.12 and 7.13. It must be noted that gages, located far from the crush zone, e.g. gages 12 and 14, measuring hoop strains, recorded initially higher values than those located near to the crush zone (gage 3); compare the relevant curves in Figure 7.13.

As progressive crushing began (loading stage 2), the strain / displacement curve for all gages follows almost the same trend as the load-deflection curve, see Figures 7.14(a) and (b) and Figure 7.15. Large strain fluctuations were recorded from the gages located very close to the crush zone (gages 3 and 4), whereas the strain gages far from the crush zone (gages 12–15) gave more stable strain readings. In addition, gage 3, after unloading, developed compressive residual strain, whilst gages far from the crush zone returned to zero strains after unloading. Note that in the stable progressive crush (serrated linear plot), there was some noticeable creep occurring in the specimen, see Figure 7.16; despite having an approximately constant load, strain readings were increasing.

From the 45 degree rosettes the directions of the principal stresses were determined to be along the longitudinal and circumferential directions; see Figure 7.13.

The length of the crack, which propagates through the resin matrix was also detected by gages 3 and 4, see Figure 7.17. At a distance approximately 7 mm before the crush zone reaches the strain gages 3 and 4, strain readings were already fluctuating and disturbed due to the effect of the crack tip approaching the gage location. Comparing with Figure 7.10(a), where the crack tip was measured to be about 7.5 mm from the surface of the crack zone, it is concluded that the results from both the strain-gage and the microscopic observations are in a very good agreement.

(b) Transition Mode of Failure (Mode II)

Circular frusta with semiapical angles greater than 20° failed following the transition mode of failure, Mode II. The strain-gages placement, for obtaining the strain attributes during collapse, is indicated in Figures 7.18 and 7.19 for the two loading stages, see also Reference [110].

The strain readings, obtained during the loading stage 1, are presented in Figure 7.20. The following remarks regarding the whole behaviour of the tested material may be drawn:

- As expected, all gages measuring strain in the axial direction showed compressive straining, whilst those measuring in the hoop direction were in tension.
- The top gages (3, 4, 7, 8) showed a rise in tensile and compressive strains as the crush zone approached them. This rise in strain occurred when the crush zone was about 25 mm from the gages. Then, the top transverse gages started to measure a decreasing tensile strain, probably due to the inner frond formation, inducing, in this manner, a compressive hoop stress.
- The axial gages (2, 4, 6, 8) showed an increase compressive strain, probably due to the radius that the inner frond is deforming through.
- The low relatively constant strain readings from the bottom gages showed that they were placed above the critical zone. This fact was also verified after the specimen was crushed. Therefore, in stage 2, the bottom gages needed to be placed closer to the base of the specimen.

From the data obtained in the loading stage 2, see Figures 7.21–7.22, the following observations may be made :

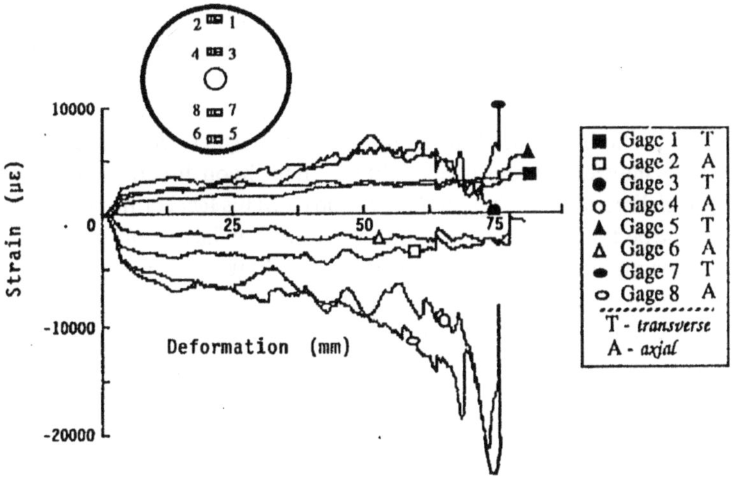

Figure 7.20. Strain distributions during stage I of loading.

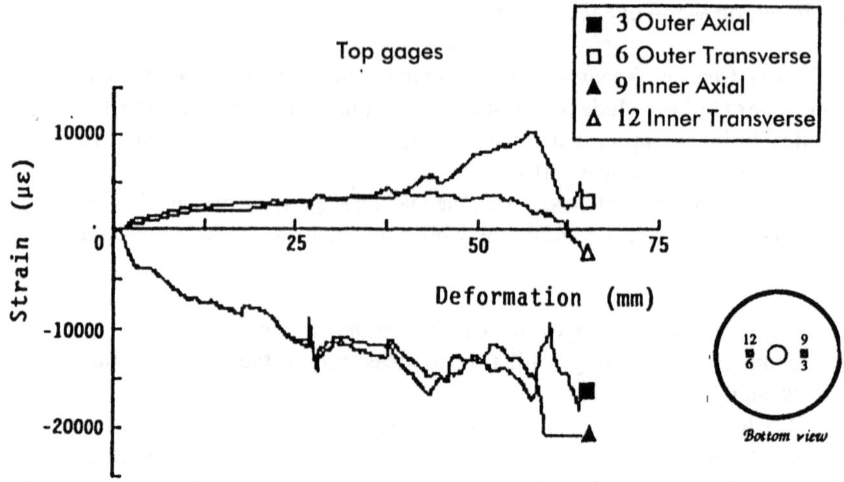

Figure 7.21. Strain distributions obtained by the top gages during stage II of loading.

- The transverse top gages 1 and 2 showed a steady equal tensile strain until the crush zone approched to about 38 mm, see Figure 7.21, and then began to bulge causing bending through its thickness. This effect is more apparent in gages 6 and 12, where the outer gage 6 showed an increase in tensile strain and the inner 12 a decrease. As the crush zone approached to about 25 mm, the same gages began to indicate decreasing tensile strains, as discussed above in stage 1.
- The axial gages 3 and 9, see Figure 7.21, showed relatively little bending, until the frond began to bend inwards. Again the outer gage 3 showed a decrease in compression strain, whilst the inner gage 9 an increase.
- As far as the bottom strain-gage data is concerned, an increase in strain is indicated, compared to the strain of the corresponding readings during the loading stage I.
- In stage I, the strains measured from the bottom gages (1, 2, 5, 6) did not exceed 2500 $\mu\varepsilon$, in both directions, see Figure 7.20, whilst, in stage 2, the corresponding readings indicated strains up to 12000 $\mu\varepsilon$ in the hoop direction and 15000 $\mu\varepsilon$ in the axial direction, see Figures 7.22(a) and (b). This fact confirmed the choice to place the bottom gages closer to the bottom edge of the frustum (critical zone).
- It is apparent that gage 10 was early lost, see Figure 7.22(a).
- The bottom axial gages 5 and 11, see Figure 7.22(a), indicated bending through the thickness of the specimen at this area. Note that, the outer gage 5 was on the compressive side of the bending and the gage 11 on the tension one.
- The transverse gages, 4 and 10 at the same location, see Figure 7.22(a), indicated also straining in bending. The outside gage 4 showed less tensile strain than the inside gage 10, when it was still giving data. The transverse gages 2 and 8, see Figure 7.22(b), showed similar offsets, supporting, in this manner, the assumption that the base of the specimen bent outwards.

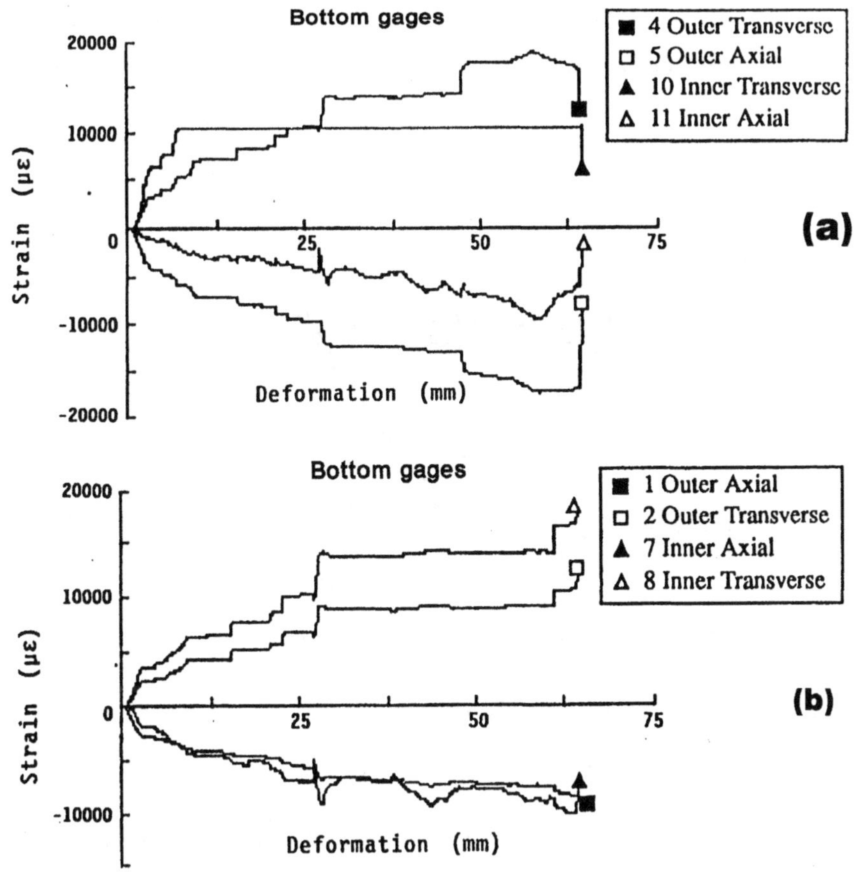

Figure 7.22. Strain distributions obtained by the bottom gages during (a) stage I of loading, (b) stage II of loading.

- The axial strain-gages 1 and 7, see Figure 7.22(b), showed negligible bending differences between the strain readings, compared to the axial strain-gages 5 and 11, located exactly opposite on the frustum, see Figure 7.22(a). It must be noted that this difference in bending around the base of the specimen indicates that the cone is not in a true state of uniform axial loading.

From the above described analysis it is evident that, during the axial compression process, the bottom of the conical frusta is in a high state of stress due to bending at their base, see Figure 7.23. This bending of the base outwards seems to be the main deformation mechanism that induces a very high tensile hoop stress at the base edge of the specimen, causing matrix cracking and leading, finally, to split-up the side wall of the conical shell according to the Mode II of collapse.

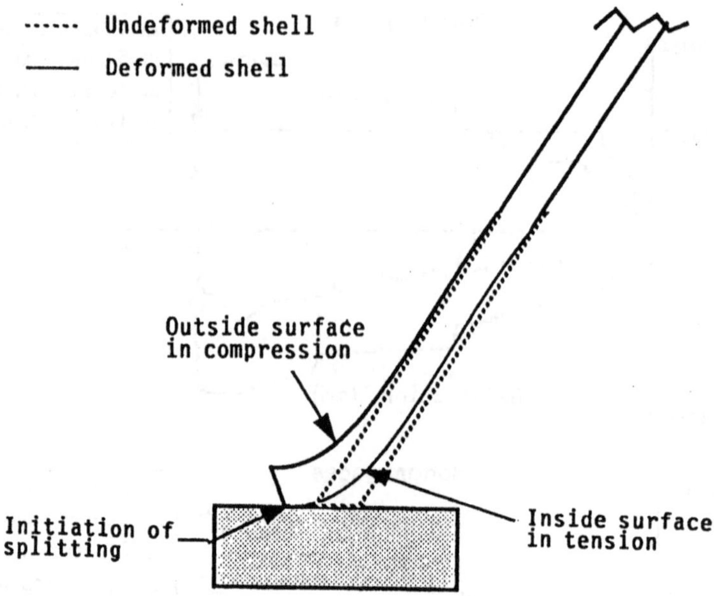

Figure 7.23. Assumed deformation at the base of an axially loaded conical frustum.

7.3.3 Energy Absorbing Characteristics

STATIC AXIAL COLLAPSE

Figures 7.1–7.7 show the variation of crushing load with decreasing axial length of the shell (the load/displacement curves) for the various collapse modes. In all cases, the shell initially behaves elastically and the load rises at a steady rate to a maximum value, P_{max}, at which time intrawall cracking forms at the top end. The initial slopes of the load/displacement curves were approximately linear and the displacement was that associated with the elastic deformation of the frustum. As deformation progressed, the rate of load increase diminished and in some cases the load dropped sharply, depending on the mode of failure.

The crushing loads and the energy absorbed were measured and tabulated in Table 7.1(a); the energy absorbed is normalised by quantifying it in terms of the energy absorbed per unit mass of material that has undergone crush. The post-crushing region of the curves is characterised by high serrations due to the crush energy of the composite material, being absorbed by a sequence of different microcracking processes characterising each case of collapse.

In the case of the end-crushing mode (Mode I), the post-crushing load/displacement curves feature three distinctive patterns. In the case of Mode Ia the post-buckling load increases approximately in linear manner, see Figure 7.1. An increasing load with displacement is expected since the effective cross-sectional area of the cone at the crush zone increases linearly. The load/displacement characteristics for

Mode Ib, see Figure 7.2, show an initial elastic behaviour and, as instability begins, the load falls off rapidly until a stable crush zone is formed. Subsequent post-crushing behaviour features a curve which is highly serrated, but the average crushing load remains almost constant. Finally, for Mode Ic, the post-crushing region shows successive fluctuations with troughs and peaks at levels lower than the initial peak load, see Figure 7.3.

The load/displacement curves for the transition mode of failure (Mode II) combine the main features of Mode I and Mode III cases, showing an initial end-crushing mode (Mode Ia or Ib) and then a sudden drop of load, see Figures 7.4 and 7.5. This mode of failure can offer an efficient mode of energy absorption, depending on the point of deformation, when transition of failure takes place. But for other specimens, see Table 7.1(a), the extent of progressive crushing was terminated at an early point of displacement, thus, showing little potential of energy absorption capability.

The load/displacement curves, corresponding to the mid-length collapse mode (Mode III), show a typical pre-crushing region, but the initial elastic response is followed by a very sharp drop in load and poor post-crushing characteristics, see Figure 7.6. Mode III is similar to that of a Euler column-buckling mode obtained when axially compressing very thin metal or PVC tubes, offering little potential for energy absorption [8].

The load-displacement curve for Mode IV shows large fluctuations and the average load is relatively small, see Figure 7.7.

Note that of the four failure modes observed, the end-crushing mode (Mode I) was found to offer the highest energy absorption capability.

The initial peak load, P_{max}, the mean post-buckling load, $\overline{P}$, and the total energy absorbed, W are greatly affected by the wall thickness, the semi-apical angle, θ and the mean diameter of the frusta, $\overline{D}$; see Table 7.1(a) for details. For constant semi-apical angle, the peak load, P_{max} increases with increasing wall thickness, whilst for constant wall thickness, the peak load decreases with increasing semi-apical angle, but increases with increasing mean diameter.

In the case of two-layered specimens, the specific energy absorbed decreases with increasing semi-apical angle, see Table 7.1(a) for details. Similarly, when collapsing three-layered specimens, the specific energy absorbed decreases at a rather moderate rate with increasing semi-apical angle from 5° to 30°.

It is evident that the energy absorbed and the mean post-crushing load increase considerably with the increase in thickness or the number of layers of glass mat. Note that, for collapsed 30° frusta, because of their early transition mode of failure from a stable progressive crushing to a separation of the wall of the frustum in two halves due to a spiral crack propagating along the shell circumference, (Mode IIb), see Figure 7.5, the energy absorbed is greatly reduced. Regardless of semi-apical angles, the mean post-crushing load increases with increasing wall thickness.

Conical frusta with large mean diameter were more prone to catastrophic failure (Mode III). The load/deflection curve, corresponding to this type of failure, see Figure 7.6, shows a pre-crushing region as a typical plot, but the onset of failure is followed by a very sharp drop in load and poor post-crushing characteristics. Therefore, the energy absorbed and the mean post-crushing load are very low.

Figure 7.24 shows the variation of crushing load with shell shortening (load-displacement curve) for both collapse modes (Modes Ia and Ib). Peak and mean crushing loads and the energy absorbed were measured and tabulated in Table 7.1(b). The energy absorbed is normalised by quantifying it in terms of the energy absorbed per unit mass of the material that has undergone crush (specific energy) and is also recorded as above.

From the shape of the dynamically obtained curves at the post-crushing region, i.e. the region after the first peak, where the main intrawall crack has already developed, it is quite difficult to make safe assumptions concerning the possible fracture mechanism occurred, as well as the development and propagation of microcracks. On the contrary, during a static test, the deformation mode encountered during collapse could be excluded from the pattern of load-deflection curve (history of deformation). Moreover, due to the dynamic nature of the phenomenon (note that duration ranges between 10–25 ms), the sequence of different microcracking processes in relation to the shape of the obtained load deflection curve can not be followed and described distinctly.

The post-crushing region of a typical dynamic load-deflection curve is characterised by a series of successive severe fluctuations with troughs and peaks, probably associated with the crush energy absorbed by the composite material through a sequence of sudden and severe microcracking processes, developed and propagating rapidly during the various phases of collapse. Note that, during the corresponding

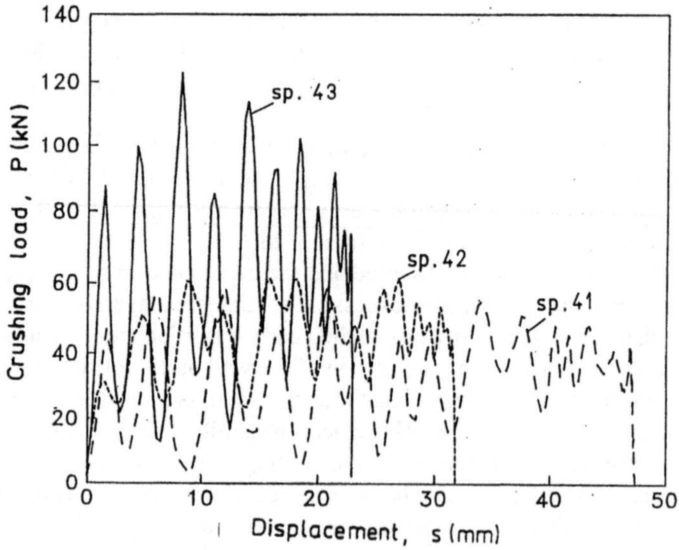

Figure 7.24. Load/deflection characteristics for dynamically loaded 15° circular frusta of various wall thicknesses (sp. 41, 42, 43; see Table 7.1).

static loading, the load/displacement curve shows a very shallow serration; compare Figs 7.1 and 7.24. The above-described characteristic feature is the main difference between the static and dynamic loading conditions. Such a phenomenon is more profound for shells with greater laminae layers (4, 5, or 6).

It is evident that the energy absorbed and the mean post-crushing load increase considerably with either the increase in thickness or the number of layers of glass mat, whilst the specific energy absorbed during crushing also varies with semi-apical angle and wall thickness. On the whole, the specific energy decreases as the semi-apical angle of the shell increases and increases significantly with the increase in thickness. There is no effect on the increase of specific energy for specimens with number of layers greater than 3; see Table 7.1(b) for details. Figure 7.25 shows how the specific energy absorbed, W_s varies with wall slenderness ratio, $t/\overline{D}$. For circular frusta is evident that with increasing $t/\overline{D}$, i.e. by increasing wall thickness or decreasing mean diameter, the specific energy absorbed increases almost linearly. Static tests for the same shell geometries develop higher values of specific energy and mean load than these obtained in dynamic testing; see Tables 7.1(a) and (b). The experimental data obtained from the static and the dynamic tests, concerning specific energy for circular frusta, fit linear equations, the plots of which are almost parallel to each other, namely

$$W_s = 205(t/\overline{D}) + 20 \tag{7.1}$$

In the same figure, plots of specific energy versus slenderness ratio, $t/\overline{D}$ ($\overline{D}$ equals the outside diameter of the tube, D), for tubes (conical shells with 0° semi-apical an-

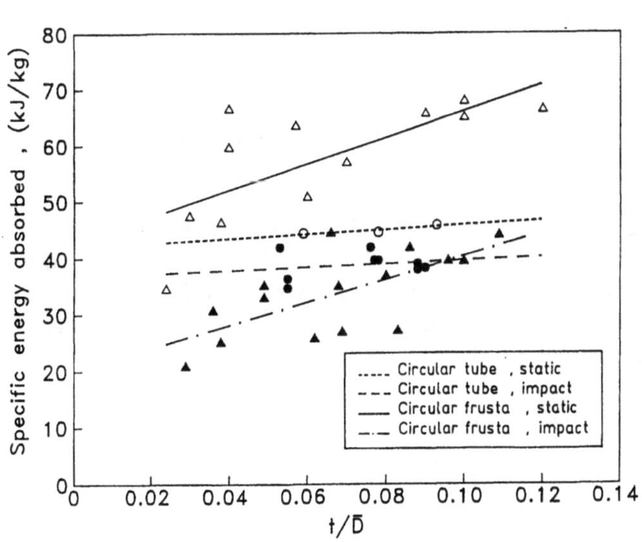

Figure 7.25 Variation of specific energy absorbed with ratio $t/\overline{D}$ (O—circular tubes static; ●—circular tubes impact; △—circular frusta static; ▲—circular frusta impact).

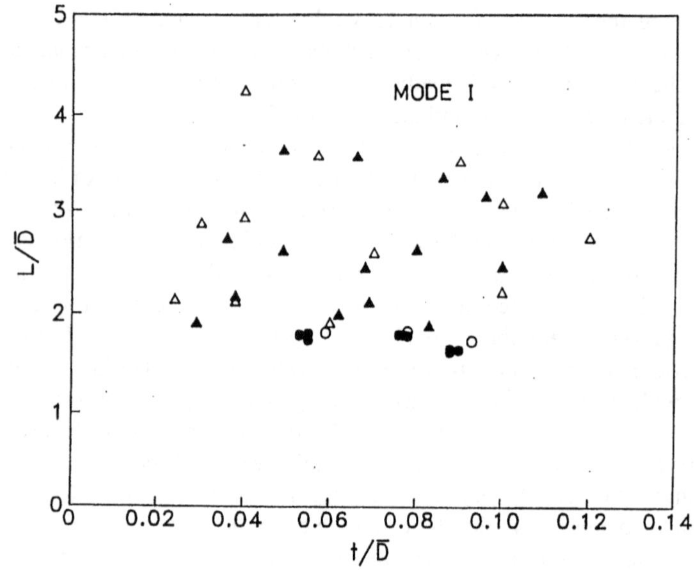

Figure 7.26. Classification chart showing the area of stable collapse mode (Mode I) for all the specimens tested; (O—circular tubes static; ●—circular tubes impact; △—circular frusta static; ▲—circular frusta impact).

gle) subjected to axial and dynamic loading, are presented. The experimental data, taken from Tables 5.1(a) and (b) of Chapter 5, for static and dynamic loading, respectively, fit the following equation

$$W_s = 236(t/\overline{D}) + 43 \tag{7.2}$$

Classifying the geometries tested from the energy absorbing capacity point of view, the geometry of a 5° circular frustum seems to be the most efficient; see Table 7.1.

In Figure 7.26, the stable mode of collapse (Mode I), obtained when axially collapsing tubes and frusta, see Tables 5.1 and 7.1, are classified with respect to the ratios $t/\overline{D}$ and $L/\overline{D}$; It is clearly shown that shell instability is prevented by ensuring that the $t/\overline{D}$ and $L/\overline{D}$ ratios are above a critical value, respectively, for the shell geometries examined. As shown in Figure 7.26, the critical value for $t/\overline{D}$ is about 0.02 and for $L/\overline{D}$ is 1.5. These critical values are almost identical for static and dynamic tests.

7.3.4 Failure Analysis

STATIC AXIAL COLLAPSE

The microfracture mechanism concerning the stable collapse mode (Mode I) of the circular frusta occurring during axial loading, is discussed in detail in the previous Sections. Two different modes of failure were observed, related to the above mentioned progressive crushing mode; the Mode Ia of failure is characterised by progressive collapse through the formation of continuous fronds which spread outwards and inwards, whilst the Mode Ib is associated with successive shearing of the region near the narrow end of the shell, accompanied by a number of delaminations and longitudinal cracks, whilst the tube wall inverses inwards, see Figure 7.9. This failure mechanism was analysed in the case of circular and square tubes in Chapters 5 and 6, respectively. The theoretical model proposed for the analysis of circular tubes was modified and is used here to analyse the collapse mechanism and to estimate the related energy absorbed during the axial crushing of conical shells.

During the elastic deformation of the shell the load rises at a steady rate to a peak value, P_{max}. At this stage a main circumferential intrawall crack of height L_c and length $L_c/\cos\theta$, forms at the top-end parallel to the axis of the shell wall, see the schematic diagram in Figure 7.10(a) and the micrograph of Figure 7.10(b), leading to a sharp drop of the load; the related shell shortening is s_1, see Figure 7.1(b). It must be noted that the position of the occurrence of the main intrawall crack moves away from the shell wall axis towards the outside edge of the shell wall with increasing semi-apical angle of the frustum, θ, whilst the crack height, L_c, decreases; see Figure 7.10. The frustum diameter at the crack tip, d_c, related to the position of the crack initiation, as measured experimentally, is

$$d_c = (9 \cdot t \cdot \theta/\pi) + d - t \tag{7.3}$$

where, following the Notation, t is the wall thickness and d the outside top diameter of the frustum. Therefore, the associated absorbed energy, which equals the external work, as obtained by measuring the area under the load/displacement curve in the elastic regime in Figure 7.1(b), is

$$\begin{aligned} W_{Lc} &= [\pi \cdot (L_c/\cos\theta) \cdot (d_c + L_c \cdot \tan\theta)] \cdot R_{ad} = \int_0^{s_1} P ds \\ &= \frac{1}{2} P_{max} \cdot s_1 \end{aligned} \tag{7.4}$$

where, R_{ad} is the fracture energy required to fracture a unit area of the adhesive at the interface between two adjacent layers.

As deformation proceeds further, the externally formed fronds curl downwards with the simultaneous development, along the circumference of the shell, of a number of axial splits, followed by splaying of the material strips, see Figure 7.9. Axial tears are not apparent in the internal fronds, which are more continuous than their external counterparts. The post-crushing regime is characterised by the formation of two lamina bundles, bent inwards and outwards due to the flexural damage; they withstand the applied load and buckle when the load or the length of the lamina bundle reaches a critical value. At this stage, a triangular debris wedge of pulverised material starts to form, see Figures 7.10 (a)–(f); its formation may be attributed to the friction between the bent bundles and the platen of the press or the drop mass of the hammer.

The energy required for the deformation mechanism, associated with the complete formation of the crush zone, equals the external work absorbed by the deforming shell in this regime, i.e.

$$W_{tr} = [\int_0^{\alpha_1+\theta} \sigma_o \cdot l_{s1} \cdot (l_{s1}/2) d\alpha + \int_0^{\alpha_2-\theta} \sigma_o \cdot l_{s2} \cdot (l_{s2}/2) d\alpha] \cdot \pi \cdot (d_c + s \cdot \tan\theta)$$
$$= \int_{s_1}^{s_2} P ds \tag{7.5}$$

where, according to the notation, σ_o is the normal stress applied by the wedge to the fronds, l_{s1} the side length of the wedge inscribed to the external bent frond, l_{s2} the side length of the wedge inscribed to the internal bent frond, α_1 the angle formed by the height and the external side of the wedge and α_2 the angle formed by the height and the internal side of the wedge, see Figure 7.10(a); s_2 is the related shell shortening corresponding to the completion of the wedge formation, see Figure 7.1(b).

Subsequent loading results in crushing with formation of internal and external fronds, whilst normal stresses develop on the sides of the debris wedge followed by shear stresses along the same sides due to the friction at the interface between the wedge and the fronds. Note, also, that additional normal and shear stresses develop at the interface between the steel press platen or the drop mass of the hammer and the deforming shell, as the formed fronds slide along this interface. The coefficient of friction in these regions depends on various phenomena associated with the material flow, such as:

- Elastic and/or plastic deformation, occurring at the contact area of the sliding surfaces subjected to external loading. In general, plastic deformation results in a decrease of the coefficient of friction due to the reduction of the shear resistance at the sliding surfaces
- Interfacial bonding due to the electrostatic forces developed in the contact area; it is greatly affected by the conductivity of the materials and the temperature field developed, leading to an increase of the friction coefficient
- Adhesion, occurring at the contact region during the sliding of the two deformable bodies of the same material, and resulting in an increase of the coefficient of friction

The crushing load in the post-crushing region is characterised by oscillations about a mean post-crushing load, $\overline{P}$, see Figure 7.1 (b); these oscillations start as soon as the formation of the debris wedge is completed. Note that, in the case of Mode Ia and for large semi-apical angles, the post-crushing load increases approximately in a linear manner. An increasing load with displacement is expected, since the effective cross-sectional area of the cone at the crush zone increases linearly; see also References [83, 11].

The total energy dissipated for the deformation of the shell for a crush distance s is given as

$$W_T = \int_{s_1}^{s} P ds \tag{7.6}$$

where, P is the average total force applied by the platen to the shell.

Since the intrawall crack propagates at a constant speed equal to the speed of the crosshead of the press, it may be assumed that the crack height, L_c remains constant. Also, the height of the wedge, h equals $0.8 \cdot (t/\cos\theta)$, as measured experimentally, see Figures 7.10(a) and (b). Therefore, taking into account the failure mechanism outlined above and, assuming that Coulomb friction prevails between the debris wedge and fronds and between fronds and platen, the dissipated energy due to friction at these interfaces for a crush distance, s can be estimated as

$$W_i = [\mu_{s1} \cdot (P_1 + P_2) + \mu_{s2} \cdot (P_3 + P_4)] \cdot \pi \cdot [d_c + (s - s_1) \cdot \tan\theta] \cdot (s - s_2) \tag{7.7}$$

where, P_1 and P_2 are the normal forces per unit length applied by the platen to the internal and external fronds, respectively, which result from static equilibrium, P_3 and P_4 are the normal forces per unit length applied to the internal and external sides of the wedge, respectively, μ_{s1} is the coefficient of friction between frond and platen and μ_{s2} the coefficient of friction between the wedge and the fronds.

Static equilibrium at the interface yields to

$$P = (P_w + P_1 + P_2) \cdot \pi \cdot [d_c + (s - s_1) \cdot \tan\theta] \tag{7.8}$$

where, P_w is the normal force per unit length applied by the platen to the debris wedge. This force is equilibrated by normal and frictional components at the uppermost frond surfaces, see Figure 7.10(a),

$$P_w = P_3 \cdot (\sin\alpha_1 + \mu_{s2} \cdot \cos\alpha_1) + P_4 \cdot (\sin\alpha_2 + \mu_{s2} \cdot \cos\alpha_2) \tag{7.9}$$

therefore,

$$P_1 + P_2 = W_T / \pi \cdot [d_c + (s - s_1) \cdot \tan\theta] \cdot s - [P_3 \cdot (\sin\alpha_1 + \mu_{s2} \cdot \cos\alpha_1) + P_4 \cdot (\sin\alpha_2 + \mu_{s2} \cdot \cos\alpha_2)] \tag{7.10}$$

Note that

$$P_3 = \sigma_o \cdot l_{s1} \tag{7.11}$$

$$P_4 = \sigma_o \cdot l_{s2} \tag{7.12}$$

$$\sigma_o = k \cdot \sigma_\theta \tag{7.13}$$

and substituting into Equation (7.7)

$$\begin{aligned} W_{\mathrm{i}} = \{\mu_{s1} \cdot [W_T / \pi \cdot [d_c + (s - s_1) \cdot \tan\theta] \cdot s - (0.8 \cdot t \cdot k \cdot \sigma_\theta / \cos\theta) \\ \cdot (\tan\alpha_1 + \tan\alpha_2 + 2 \cdot \mu_{s2})] + \mu_{s2} \cdot (0.8 \cdot t \cdot k \cdot \sigma_\theta / \cos\theta) \\ \cdot (1/\cos\alpha_1 + 1/\cos\alpha_2)\} \cdot \pi \cdot [d_c + (s - s_1) \cdot \tan\theta] \cdot (s - s_2) \end{aligned} \tag{7.14}$$

where, k is a constant and σ_θ the tensile fracture stress of the composite material, see Figure 7.10(a).

The energy dissipated due to fronds bending, causing fibre debonding, delaminations and other failures, can be estimated as

$$\begin{aligned} W_{\mathrm{ii}} &= \{ \int_0^{\alpha_1+\theta} P_3 \cdot (l_{s1}/2) d\alpha + \int_0^{\alpha_2-\theta} P_4 \cdot (l_{s2}/2) d\alpha \\ &+ \int_{s_2}^{s} [P_3 \cdot (\alpha_1 + \theta) + P_4 \cdot (\alpha_2 - \theta)] ds\} \cdot \pi \cdot [d_c + (s - s_1) \cdot \tan\theta] \\ &= (0.8 \cdot t \cdot k \cdot \sigma_\theta \cdot \pi \cdot [d_c + (s - s_1) \cdot \tan\theta] / \cos\theta) \\ &\cdot \{[(\alpha_1 + \theta) / \cos\alpha_1)] \cdot [0.4 \cdot t/(\cos\alpha_1 \cdot \cos\theta) + s - s_2] \\ &+ [(\alpha_2 - \theta) / \cos\alpha_2)] \cdot [0.4 \cdot t/(\cos\alpha_2 \cdot \cos\theta) + s - s_2]\} \end{aligned} \tag{7.15}$$

whilst the energy dissipated due to crack growth is expressed as

$$\begin{aligned} W_{\mathrm{iii}} = R_{\mathrm{ad}} \cdot [(s - s_1) \cdot \pi \cdot [d_c + (s - s_1) \cdot \tan\theta] \\ + \pi \cdot (L_c / \cos\theta) \cdot (d_c + L_c \cdot \tan\theta)] \end{aligned} \tag{7.16}$$

Axial splits are formed due to tensile stresses developed in the hoop direction of the shell, resulting in the formation of external fronds, see Figures 7.9 and 7.10; the energy absorbed by these fronds, as they curl downwards with subsequent splaying of material strips, is

$$W_{\mathrm{iv}} = n \cdot (t/2 - 4.5 \cdot \theta \cdot t/\pi) \cdot G \cdot s \tag{7.17}$$

where, n is the number of axial splits and G the fracture toughness. The number of axial splits, n can be determined from the ratio

$$n = 2 \cdot \pi / b_{cr} \tag{7.18}$$

where, b_{cr} is the angle between two splits, see Figure 7.9.

The total energy dissipated for the deformation of the shell is given as

$$W_T = W_{\mathrm{i}} + W_{\mathrm{ii}} + W_{\mathrm{iii}} + W_{\mathrm{iv}} \tag{7.19}$$

and substituting Equations (7.14–7.17) into Equation (7.19)

$$\begin{aligned} W_T &= [1/(1-\mu_{s1}+\mu_{s1}\cdot s_2/s)]\cdot\{[0.8\cdot t\cdot k\cdot\sigma_\theta\cdot\pi\cdot[d_c+(s-s_1)\cdot\tan\theta]/\cos\theta] \\ &\cdot[(s-s_2)\cdot[(\mu_{s2}\cdot(1/\cos\alpha_1+1/\cos\alpha_2)-\mu_{s1}\cdot(\tan\alpha_1+\tan\alpha_2 \\ &+2\cdot\mu_{s2})]+[(\alpha_1+\theta)/\cos\alpha_1]\cdot[0.4\cdot t/(\cos\alpha_1\cdot\cos\theta)+s-s_2] \\ &+[(\alpha_2-\theta)/\cos\alpha_2]\cdot[0.4\cdot t/(\cos\alpha_2\cdot\cos\theta)+s-s_2]] \\ &+R_{ad}\cdot[(s-s_1)\cdot\pi\cdot[d_c+(s-s_1)\cdot\tan\theta]+\pi\cdot(L_c/\cos\theta) \\ &\cdot(d_c+L_c\cdot\tan\theta)]+n\cdot(t/2-4.5\cdot\theta\cdot t/\pi)\cdot G\cdot s\} \end{aligned} \tag{7.20}$$

Conical shells of larger semi-apical angles $\theta \geq 15°$ collapse following the Mode Ib of failure. The total energy absorbed in this case is calculated

$$W_T = W_{\text{i}} + W_{\text{ii}} + W_{\text{iii}} \tag{7.21}$$

where,

$$W_{\text{i}} = \mu_{s1}\cdot P\cdot s = \mu_{s1}\cdot W_T \tag{7.22}$$

$$W_{\text{ii}} = P\cdot\sin\theta\cdot(\pi/2-\theta)\cdot s = \sin\theta\cdot(\pi/2-\theta)\cdot W_T \tag{7.23}$$

$$W_{\text{iii}} = R_{\text{ad}}\cdot\pi\cdot s\cdot[(d-t)+s\cdot\tan\theta]\cdot(N-1) \tag{7.24}$$

and substituting Equations (7.22–7.24) into Equation (7.21)

$$W_T = R_{\text{ad}}\cdot\pi\cdot s\cdot[(d-t)+s\cdot\tan\theta]\cdot(N-1)/[1-\mu_{s1}-\sin\theta\cdot(\pi/2-\theta)] \tag{7.25}$$

where, N is the number of layers of the shell. Note that the energy dissipated due to axial splits, W_{iv} equals zero.

THE EFFECT OF STRAIN-RATE

As reported in Section 7.3.2, the microfracture mechanism for the progressive collapse of conical shells subjected to dynamic loading is, in general, similar to that obtained during the axial static collapse. The only differences encountered are dealing with the shape and the position of the wedge and the microcracking development; compare Figures 7.10(f) and 7.10(i). Therefore, the above proposed theoretical model can be applied for the prediction of the crushing loads and the energy absorbed in the case of circular frusta subjected to axial dynamic loading.

The effect of strain-rate on the crushing behaviour of axially loaded composite tubes is reviewed in References [69, 82]. Strain-rate and, therefore, crush-speed may influence the mechanical properties of the fibre and the matrix of the composite material. Brittle fibres are, in general, insensitive to strain-rate, whilst the fracturing of the lamina bundles is not a function of crush-speed. Strain-rate has an effect on matrix stiffness and failure strain and, therefore, the energy absorption, associated with the interlaminar crack growth, may be considered as a function of the crush-speed [69]. Note that the coefficient of friction between the composite material and the crushing surface and between the debris wedge and the fronds and, subsequently, the energy absorption capability of the shell are also affected from the crush speed [67, 82].

The major part of the absorbed energy, during the static axial compression of a shell, is dissipated as frictional work in the crushed material or at the interface between the material and the tool and this is estimated to be about 50% or more of the total work done; see also References [75, 93]. In dynamic collapse, attention is directed towards the influence of strain-rate on the frictional work absorbed during the impact, taking into account all structural and material parameters, which may contribute to it, i.e. the fibre and matrix material, the fibre diameter and orientation in the laminate, the fibre volume content as well as the conditions at the interfaces.

The amount of energy dissipated due to fronds bending, W_{ii}, to crack growth, W_{iii} and to axial splits, W_{iv} are estimated from Equations (7.15), (7.16) and (7.17), respectively, whilst the energy dissipated due to friction, W_i is estimated from Equation (7.14) by substituting the static coefficients, μ_{s1} and μ_{s2} with the related dynamic ones, μ_{d1} and μ_{d2}, respectively. Therefore, the total energy absorbed during the dynamic crushing of the shells can be calculated from Equation (7.20), properly modified, as

$$\begin{aligned} W_T = {} & [1/(1-\mu_{d1}+\mu_{d1}\cdot s_2/s)]\cdot\{[0.8\cdot t\cdot k\cdot\sigma_\theta\cdot\pi\cdot[d_c+(s-s_1)\cdot\tan\theta]/\cos\theta] \\ & \cdot[(s-s_2)\cdot[(\mu_{d2}\cdot(1/\cos\alpha_1+1/\cos\alpha_2)-\mu_{d1}\cdot(\tan\alpha_1+\tan\alpha_2 \\ & +2\cdot\mu_{d2})]+[(\alpha_1+\theta)/\cos\alpha_1]\cdot[0.4\cdot t/(\cos\alpha_1\cdot\cos\theta)+s-s_2] \\ & +[(\alpha_2-\theta)/\cos\alpha_2]\cdot[0.4\cdot t/(\cos\alpha_2\cdot\cos\theta)+s-s_2]] \\ & +R_{ad}\cdot[(s-s_1)\cdot\pi\cdot(d_c+(s-s_1)\cdot\tan\theta)+\pi\cdot(L_c/\cos\theta) \\ & \cdot(d_c+L_c\cdot\tan\theta)]+n\cdot(t/2-4s\cdot\theta\cdot t/\pi)\cdot G\cdot s\} \end{aligned} \tag{7.26}$$

where, μ_{d1} and μ_{d2} are the corresponding dynamic coefficients of friction.

The total energy absorbed in the case of the dynamic collapse of conical shells, following the Mode Ib of failure, is obtained by modifying Equation (7.25) as

$$W_T = R_{ad}\cdot\pi\cdot s[(d-t)+s\cdot\tan\theta]\cdot(N-1)/[1-\mu_{d1}-\sin\theta\cdot(\pi/2-\theta)] \tag{7.27}$$

Note that, in the case of dynamic loading, transition from Mode Ia to Mode Ib is observed for smaller semi-apical angles, θ, as compared to the static ones; see Table 7.1 and also Reference [82].

From the stress / strain curve for the composite material A, shown in Figure 5.4 of Chapter 5, the tensile fracture stress, σ_θ of the material was estimated. The static friction coefficients, μ_{s1} and μ_{s2} were obtained by employing the curling test [41]. An estimate of the dynamic coefficients of friction at the fronds/platen interface, μ_{d1} and the fronds/wedge interface, μ_{d2} was obtained by comparing the experimental crushing loads in Table 7.1. The fracture toughness, G was estimated from the tension test of notched strips. The interfacial fracture energy, R_{ad} was calculated from Equation (5.1) of Chapter 5 using the experimental results obtained by loading cylindrical tubes up to the maximum load, P_{max} and obtaining the energy absorbed from the related experimental load/displacement curves of loaded shells. Note, however, that the R_{ad} may be also estimated by employing the strip peel test, see Reference [46]. Experimentally obtained values for $\mu_{s1}, \mu_{s2}, \mu_{d1}, \mu_{d2}, R_{ad}, G, \sigma_\theta$ and the constant k are presented in Table 5.2 of Chapter 5.

7.3.5 Crashworthy Capability: Concluding Remarks

The crush behaviour and the energy absorbing characteristics of fibre-reinforced circular frusta, subjected to static and dynamic axial collapse, was examined both theoretically and experimentally. Detailed macro- and microstructural investigations on a variety of crushed conical shells gave useful information and contributed greatly to the understanding of the processes involved.

Four modes of collapse at macroscopic level were observed:

- an end-crushing mode (Mode I) characterised by progressive collapse, starting at one end of the shell
- a transition mode of collapse (Mode II), showing successive formations of an end-crushing mode and a circumferential crack at a just lower position
- a mid-length collapse (Mode III), featuring a brittle fracture involving catastrophic failure
- a progressive folding mode (Mode IV), with the formation of sharp hinges. Conical frusta can dissipate large amounts of energy by stable collapse (Mode I), when subjected to axial loading. The transition point between stable and unstable mode of collapse, due to the effect of the frustum semi-apical angle, was identified to be between 15° and 20° of semi-apical angle

The transitional crushing phenomena, the formation and the growth of the collapse mechanism, concerning the axial collapse of circular frusta of fibreglass composite materials, are complicated and stochastic, however, the proposed theoretical analysis was experimentally verified. The mean post-crushing load, $\overline{P}$, the total energy absorbed, W_T and the specific energy, W_s are well predicted theoretically by the proposed analysis within $\pm 15\%$, see Table 7.1. Higher energies and loads were predicted for shells with even number of layers, see Table 7.1; however, it must be noted that fibre orientation was not taken into account. For the prediction of the crushing loads and the absorbed energy of the dynamically loaded shells, it was assumed that, the tensile fracture stress and the fracture toughness of the material remain constant, whilst these parameters can be a function of the crush-speed.

As outlined above, the tip of the main intrawall crack is shifted towards the outer edge of the shell wall as the frusta semi-apical angle, θ increases, whilst the size of the wedge, as well as the length of the main crack, L_c, reduce, due to the transition from Mode Ia to Mode Ib, see Figure 7.10. A great amount of the absorbed energy is dissipated for the deformation and fracture of the internal and external fronds as they are bent over. The radius of the internal frond, r_i increases with increasing semi-apical angle (the angle bent over is $\alpha_2 - \theta$), whilst the external frond is constrained to deform through a smaller radius, r_o (the angle bent over is $\alpha_2 + \theta$). The external frond becomes smaller as the semi-apical angle increases, although bending through a greater angle occurs as θ increases. Contrariwise, the internal frond becomes larger than the external one, but it is bent through a smaller angle as θ increases. Subsequently, the specific energy, W_s, predicted both theoretically and experimentally, decreases as the semi-apical angle increases, see also Table 7.1. Therefore, the amount of energy dissipated to forming internal and external fronds due to bending depends on various parameters, such as the magnitudes of angles, α_1 and α_2, the semi-apical angle, θ, and the position of the main crack tip.

In general, the microfracture mechanism of Mode Ia of collapse is similar for statically and dynamically loaded shells. The only differences encountered are associated with the shape of the pulverised wedge and the microcracking development.

In the case of the dynamically loaded circular frusta, the main cracking caused by the wedge is smaller in size, whilst the position of the wedge, which is smaller in size, and of the tip of the crack are shifted towards the outside wall surface of the impacted shells, as compared to statically loaded ones, Therefore, transition from Mode Ia to Mode Ib occurs for smaller semi-apical angles for dynamically loaded frusta, as compared to the related static ones.

Lower values of the energy absorbed, by about 20%, were obtained for dynamically loaded shells, as compared with those predicted in static collapse, propably due to the lower values of the dynamic friction coefficients between the wedge/fronds and fronds/platen interface and to the effect of the strain-rate on the microfracture mechanism, outlined above.

As far as the Mode Ia of collapse of the conical shells is concerned, from the proposed analysis the distribution of the dissipated energy (average results for all the semi-apical angles used) of the crushed shell, associated with the main energy sources can be estimated as: Energy absorbed, due to friction between the annular wedge and the fronds and between the fronds and the platen about 48%; due to bending of the fronds about 44%; due to crack propagation about 6%; due to axial splitting about 2%. Note that, for the dynamically loaded circular tubes, the related amounts of the energy dissipated due to friction, fronds bending, crack growth and axial splits were estimated to about 50%, 40%, 7% and 3%, respectively. It is, therefore, evident that the frictional conditions between wedge/fronds and fronds/platen constitute the most significant factors to the energy absorbing capability of the shell; the friction coefficients, μ_{s1}, μ_{d1} and μ_{s2}, μ_{d2} are greatly affected by the surface conditions at the interfaces between composite material/platen or drop-mass and composite material/debris wedge, respectively.

In the case of circular frusta following the Mode Ib of collapse, the distribution of the energy absorbed due to friction, fronds bending and delaminations was estimated to about 30%, 45% and 25%.

From the energy absorbing capability point of view, circular frusta of 5° semi-apical angle seem to be more efficient, due to the fact that an optimum combination of high values of energy dissipated both for frictional work and fronds bending was obtained and verified theoretically. The microfracture mechanism of this shell geometry is very similar to that of circular tubes, see also the similar observations reported in Reference [91].

CHAPTER 8

SQUARE FRUSTA

8.1 NOTATION

b = side width of square frustum
b_e = side width of square frustum at crack tip
C = mean circumference of square frustum
G = fracture toughness
h = height of wedge
k = constant
L = axial length of square frustum
L_c = height of central crack
l_s = side length of wedge
n = number of axial splits
P = current crushing load
P_1, P_2 = normal force per unit length
$\overline{P}$ = mean crushing load
P_{max} = peak load
R_{ad} = fracture energy per unit area of layers
r = radius of curvature of the frond
s = displacement, shell shortening, crush length
t = wall thickness of square frustum
v = crush-speed
W = energy absorbed
W_T = total energy dissipated
W_s = specific energy
W_{tr} = energy required for the crush zone formation
α = angle of wedge
ε = strain
θ = semi-apical angle of square frustum
μ_s = static friction coefficient

μ_d = dynamic friction coefficient
σ_θ = tensile fracture stress
σ_o = normal stress

8.2 GENERAL

The crashworthy behaviour of square frusta of fibreglass composite material, subjected to axial loading at various strain-rates, is reported in this chapter, see also Reference [92]. The effect of specimen geometry and the loading rate on the energy absorbing capability was experimentally studied. The mechanics of the axial crumbling process, from macroscopic and microscopic point of view, were also investigated theoretically and experimentally. The collapse modes at macroscopic and microscopic scale during the failure process were observed and analysed. A theoretical analysis of the observed stable collapse mechanism of the components, crushed under axial compression, for calculating crushing loads and energy absorbed during collapse, is proposed. A good agreement between theoretical and experimental results was obtained, indicating the efficiency of the theoretical model in predicting the energy absorbing capacity of the collapsed shell.

8.3 AXIAL COLLAPSE: STATIC AND DYNAMIC

8.3.1 Experimental

The testing material was a fibreglass composite material, with individual fibre diameter of 9 μm chopped strand mat with random fibre orientation in the plane of the mat, designated as composite material A. The shells were fabricated by a hand lay-up technique using pieces of fibreglass cloth (0.8 g/mm^2) and impregnating it with a polyester resin, providing in this manner with a composite material of 72% per weight fibre content and 1.37 g/cm^3 density. Details about the fabrication of the composite shell are presented in Section 5.3.1 of Chapter 5. The stress-strain curve as obtained from quasi-static tensile testing for the material tested (material A) is shown in Figure 5.4 of Chapter 5.

Likewise of circular frusta (Chapter 7), six different square frusta forms were fabricated from laminated hard maple wooden blocks, so that a 5°, 10°, 15°, 20°, 25° and 30° semi-apical angle was attained, respectively. Details about the forms preparation are reported in Section 7.3.1 of Chapter 7.

The static axial collapse was carried out between the parallel steel platens of a SMG hydraulic press at a crosshead speed of 10 mm/min or a compression strain-rate of 10^{-3} sec^{-1}. The corresponding dynamic tests were performed by direct impact on a drop-hammer at velocities exceeding 1 m/s. The existing drop-hammer facility, with a 75 kg falling mass from a maximum drop height of 4 m, provides a maximum impact velocity of about 10 m/s. The experimental set-up and measuring devices used throughout the present tests are described in detail in Section 5.3.1 of Chapter 5.

Load/shell shortening (displacement) curves during the crushing process were automatically measured and recorded for both types of loading. The values of the initial peak load, P_{max} and the energy absorbed, W for the axially collapsed specimens, obtained by measuring the area under the load/displacement curve, as well as the mean post-crushing load, $\overline{P}$ (defined as the ratio of energy absorbed to the total shell shortening) and the specific energy, W_s (equal to the energy absorbed per unit mass crushed, calculated as the crushed volume times the density of the material) are tabulated in Table 8.1.

Photographs showing characteristic terminal views of the deformed specimens for all series of experiments are presented in Figure 8.1. Typical micrographs of the crush zone, showing the main microfailures, as obtained using a Unimet metallographic optical microscope, are also shown in Figures 8.2–8.5. For details about specimens preparation and polishing for the microscopic observations, see Section 5.3.1 of Chapter 5.

8.3.2 Failure Mechanisms: Experimental Observations

Four distinct collapse modes at macroscopic scale, designated as Mode I, II, III and IV, respectively, were observed throughout the axial static and dynamic loading of square frusta. These modes are essentially similar to those observed in the previous investigation concerning circular frusta, see Section 7.3.2 of Chapter 7. These collapse modes mainly depend upon the wall thickness (number of layers), the semi-apical angle, θ, the axial length, L and the mean circumference, C of the shell and the testing conditions.

The main characteristics of the four collapse modes observed, are outlined below:

PROGRESSIVE END-CRUSHING (MODE I)

Two different modes of failure were observed concerning the above mentioned progressive crushing mode. *Mode Ia* of failure, similar to a "mushrooming" failure, is characterised by progressive collapse through the formation of continuous fronds which spread outwards and inwards; see Figures 8.1(a) and (b) and also References [75, 92, 95]. At the stage that the crush load reaches the peak value, P_{max}, cracks form at each of the four corners, accompanied by the formation of a circumferential intrawall crack at the end of the shell adjacent to the loading area, leading to an abrupt drop of the load, see Figure 8.6(a). As deformation proceeds further, the externally formed fronds curl downwards, with the simultaneous development of four axial splits followed by splaying of the material strips, see Figures 8.1(a) and (b); note that the splits are always located at the four corners of the narrow end of the shell, probably due to local stress concentration during the very early stages of straining. Axial tears were not apparent in the internal fronds, which were more continuous than their external counterparts.

The post-crushing regime is characterised by the formation of two lamina bundles, bent inwards and outwards, due to the flexural damage, which occurs at a distance from the contact surface equal to the wall thickness; they withstand the applied load

Table 8.1: Crushing characteristics of axially loaded square frusta

(a) Static

Sp. No	Semi-apical angle, θ (°)	Number of layers	Thick-ness, t (mm)	Axial length, L (mm)	Outside width, mm		Mean circum-ference, C* (mm)	t/C	L/C	Crush length, s (mm)	Collapse Mode	Crushing load, P (kN)			Total energy absorbed, W_T (kJ)		Specific energy, Ws (kJ/kg)	
					Bottom end, b_1	Top end, b_2						Initial peak, P_{max}	Mean post-crushing, $\overline{P}$					
													Exper.	Theor.	Exper.	Theor.	Exper.	Theor.
1	5	1	1.0	165.1	60.9	32.0	185.8	0.005	0.89	59.2	IV	4.6	0.9	-	0.053	-	-	-
2	5	2	2.5	171.4	61.0	31.0	184.0	0.014	0.93	9.1	III	29.2	7.6	-	0.069	-	-	-
3	5	2	2.5	146.0	60.4	34.8	190.4	0.013	0.77	63.5	Ia	32.6	18.1	19.1	1.149	1.214	31.8	33.6
4	5	3	3.8	176.5	71.5	40.6	224.2	0.017	0.79	63.5	Ia	66.4	41.5	42.4	2.635	2.695	42.6	43.6
5	5	5	5.8	158.7	66.1	38.4	209.0	0.028	0.76	63.5	Ia	143.2	66.2	67.3	4.204	4.276	46.4	47.2
6	10	1	1.2	170.8	91.2	31.2	244.8	0.005	0.70	53.7	IV	4.4	2.7	-	0.145	-	-	-
7	10	2	2.5	174.0	112.4	51.1	327.0	0.008	0.53	15.0	III	35.7	8.2	-	0.123	-	-	-
8	10	2	2.5	127.0	80.4	35.6	232.0	0.011	0.54	6.7	III	39.6	5.9	-	0.039	-	-	-
9	10	3	4.5	163.3	95.7	38.1	267.6	0.017	0.61	63.5	Ia	49.8	39.4	41.2	2.502	2.619	39.1	40.9
10	10	5	6.3	128.5	84.9	39.6	249.0	0.025	0.51	63.5	Ia	108.5	73.8	75.4	4.686	4.788	40.7	41.6
11	15	1	1.5	163.2	128.1	40.6	337.4	0.004	0.48	63.6	IV	5.6	2.0	-	0.127	-	-	-
12	15	2	2.5	158.5	120.8	35.8	313.2	0.008	0.50	63.6	Ia	36.2	18.0	18.6	1.145	1.185	28.4	29.4
13	15	3	3.8	153.7	121.7	39.4	322.2	0.012	0.47	63.5	Ia	53.5	39.4	40.3	2.502	2.588	37.0	38.3
14	15	5	6.3	144.5	113.5	36.1	299.2	0.021	0.48	63.7	Ia	169.1	81.4	79.5	5.185	5.064	40.1	39.2
15	20	1	1.0	124.5	141.7	51.1	385.6	0.003	0.32	24.6	II	5.7	2.9	-	0.071	-	-	-
16	20	2	2.5	158.0	164.5	49.5	428.0	0.006	0.36	64.0	II	25.2	9.8	-	0.627	-	-	-
17	20	3	3.8	157.6	134.6	49.0	367.2	0.010	0.32	16.9	II	37.5	10.5	-	0.177	-	-	-
18	25	1	1.2	123.2	188.6	73.7	524.6	0.002	0.23	32.4	II	2.9	1.2	-	0.039	-	-	-
19	25	2	2.5	142.2	191.1	58.4	499.0	0.005	0.38	28.2	II	8.9	8.0	-	0.226	-	-	-
20	30	1	1.2	142.2	215.8	51.6	534.8	0.002	0.26	26.5	II	2.0	1.0	-	0.265	-	-	-
21	30	2	2.5	146.0	228.3	59.7	576.0	0.004	0.25	14.2	II	8.4	4.9	-	0.069	-	-	-

Table 8.1 (cont.)

(b) Dynamic

Sp. No	Semi-apical angle. θ (°)	Number of layers	Thick-ness, t (mm)	Axial length, L (mm)	Outside width, mm		Mean circum-ference, C* (mm)	t/C	L/C	Crush speed, v (m/s)	Crush length, s (mm)	Collapse Mode	Crushing load, P (kN)			Total energy absorbed, W_T (kJ)		Specific energy, Ws (kJ/kg)	
					Bottom end, b_1	Top end, b_2							Initial peak, P_{max}	Mean post-crushing, $\bar{P}$					
														Exper.	Theor.	Exper.	Theor.	Exper.	Theor.
22	5	2	2.2	159.0	60.8	41.7	205.2	0.011	0.78	6.0	91.5	Ia	39.4	12.5	13.7	1.106	1.250	24.5	27.7
23	5	3	3.1	158.6	67.0	41.5	217.2	0.014	0.73	6.0	41.3	Ia	57.9	26.2	29.9	1.100	1.235	37.7	42.3
24	5	4	4.2	158.5	71.8	46.1	235.2	0.018	0.68	7.0	40.9	Ia	82.8	38.7	39.9	1.584	1.630	37.2	38.3
25	5	5	5.5	161.3	72.5	47.9	240.8	0.023	0.67	7.0	28.3	Ia	128.8	53.4	57.2	1.511	1.620	39.2	42.1
26	5	6	6.3	160.0	74.2	50.8	250.0	0.025	0.64	7.0	24.9	Ia	184.7	63.2	66.1	1.572	1.645	39.3	41.1
27	10	2	2.3	145.6	95.5	46.7	284.4	0.008	0.51	6.0	66.3	Ia	16.8	16.6	18.5	1.103	1.226	23.2	25.8
28	10	3	3.2	155.3	91.0	38.3	258.8	0.012	0.60	6.0	44.7	Ia	22.0	27.0	26.9	1.209	1.200	35.4	35.3
29	10	4	4.3	149.0	97.6	46.0	287.2	0.015	0.52	7.0	37.9	Ia	34.5	39.5	41.7	1.499	1.580	35.4	37.4
30	10	5	6.2	160.4	103.5	48.3	303.6	0.021	0.53	7.0	28.4	Ia	71.5	53.0	56.1	1.504	1.593	32.8	34.7
31	10	6	7.0	156.2	106.2	54.4	321.2	0.022	0.49	7.0	20.4	Ia	70.5	73.6	78.4	1.586	1.599	37.2	37.5
32	15	2	2.6	142.5	118.3	46.2	329.2	0.008	0.43	6.0	58.1	II	14.1	11.5	-	0.669	-	-	-
33	15	3	3.6	153.7	124.9	45.8	341.6	0.011	0.45	6.0	44.1	Ia	34.9	24.9	28.9	1.100	1.275	28.5	33.0
34	15	4	4.5	145.6	125.8	49.8	351.2	0.013	0.42	7.0	38.6	Ia	47.8	38.9	41.3	1.500	1.595	30.1	32.0
35	15	5	6.0	149.0	129.7	52.2	362.8	0.017	0.41	7.0	28.6	Ib	74.0	54.0	56.1	1.542	1.604	29.4	30.6
36	15	6	7.3	160.5	132.1	49.6	363.6	0.020	0.44	7.0	23.3	Ib	99.8	63.4	68.1	1.478	1.586	32.3	34.6

* $C=2 \cdot (b_1+b_2)$

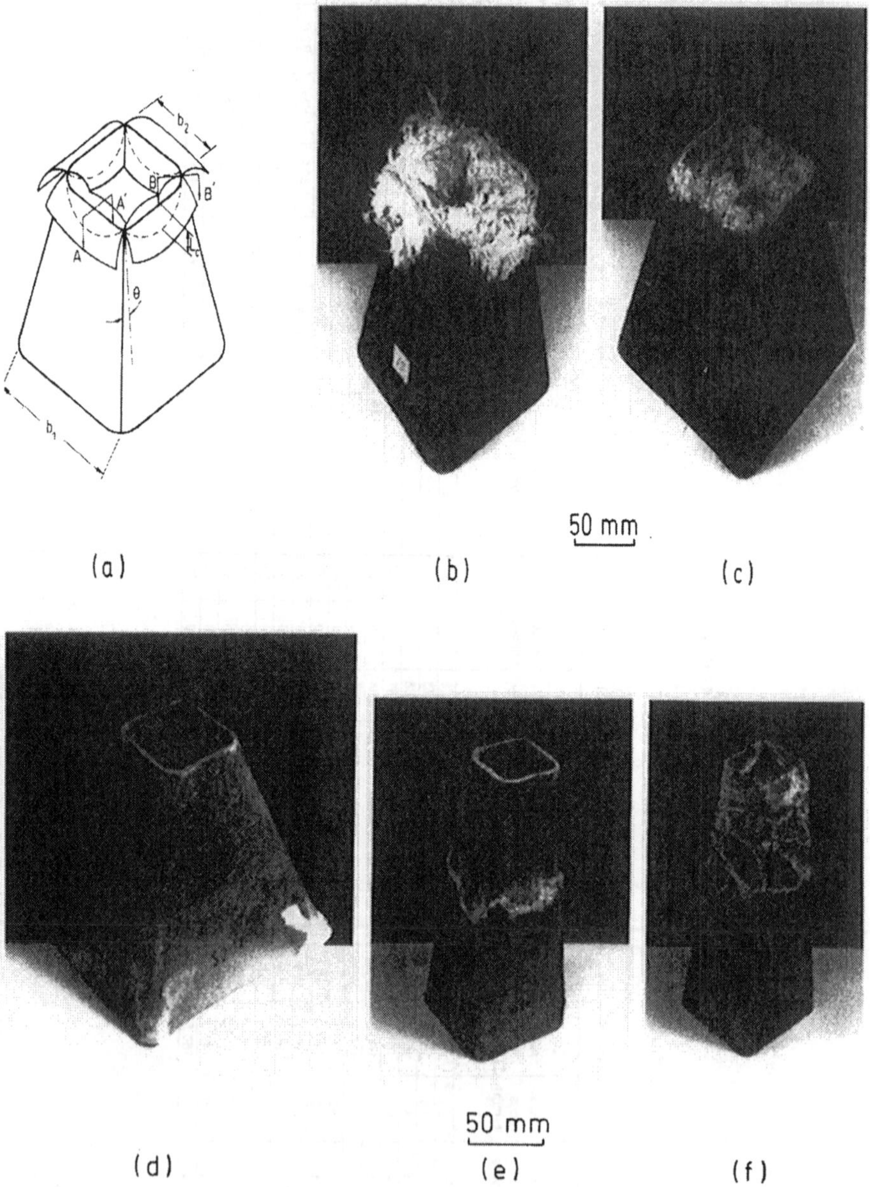

Figure 8.1. (a) Configuration of failure mechanism of Mode Ia. Macroscopic view of the various collapse modes (see Table 8.1): (b) Mode Ia (sp. 9), (c) Mode Ib (sp. 33), (d) Mode II (sp. 17), (e) Mode III (sp. 2), (f) Mode IV (sp. 1).

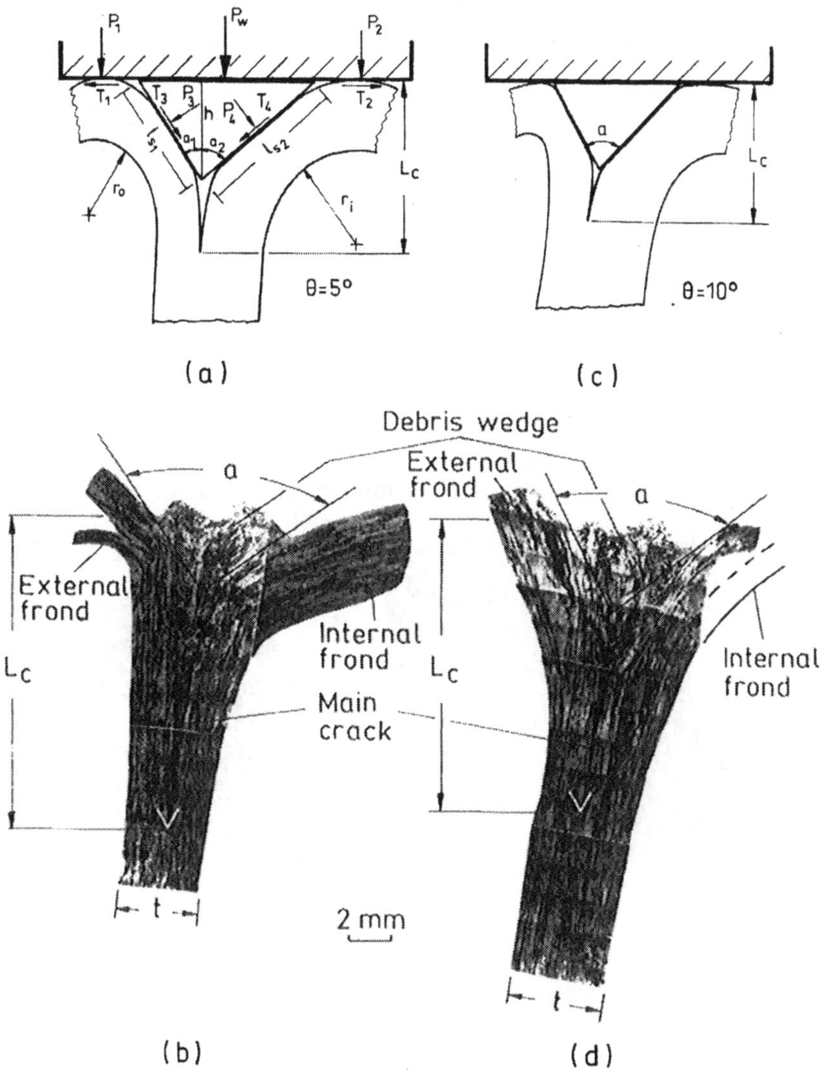

Figure 8.2. (a) Configuration of the crush zone through the wall thickness of a 5° specimen (cross section AA′ in Fig. 8.1(a)); (b) micrograph showing microfailures in the crush-zone (Mode Ia) for a statically loaded 5° square frustum (sp. 4; see Table 8.1) in a cross section at the middle of the square side; (c) configuration of the crush zone through the wall thickness of a 10° specimen (cross section AA′ in Figure 8.1(a)), (d) micrograph showing microfailures in the crush-zone (Mode Ia) for a statically loaded 10° square frustum (sp. 9, see Table 8.1) in a cross section at the middle of the square side.

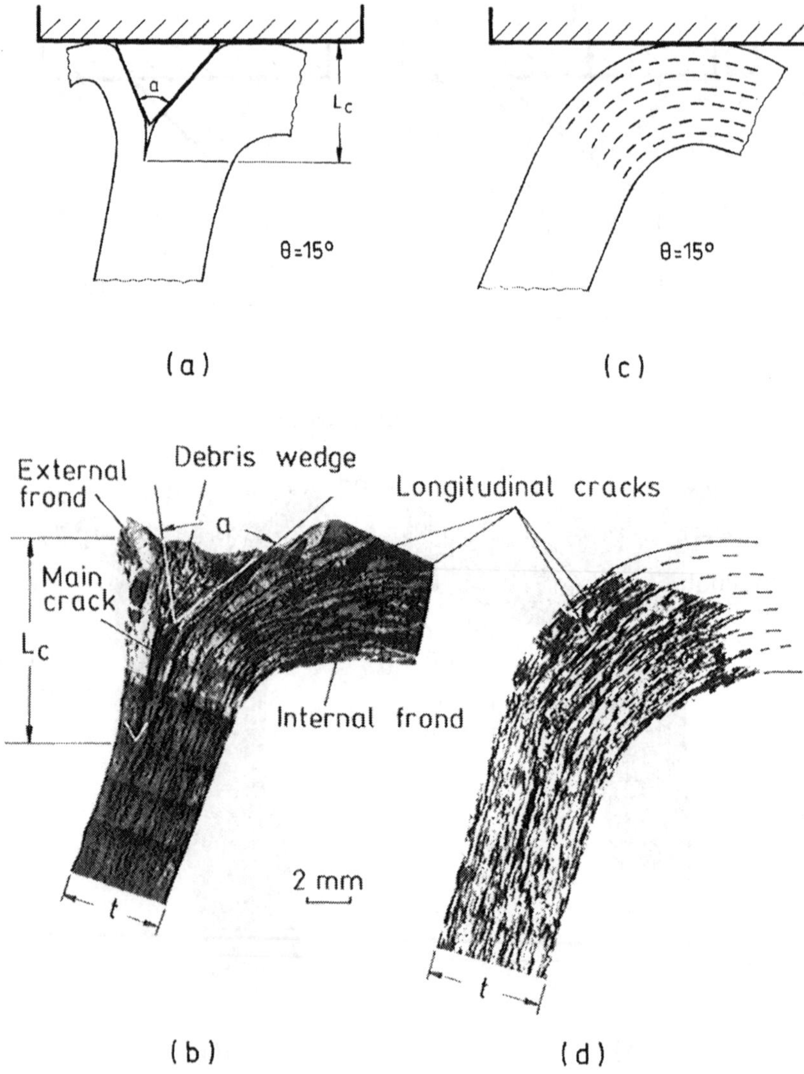

Figure 8.3. (a) Configuration of the crush-zone (Mode Ia) through the wall thickness of a 15° specimen (cross section AA′ in Figure 8.1(a)), (b) micrograph showing microfailures in the crush-zone (Mode Ia) for a statically loaded 15° square frustum (sp. 13; see Table 8.1) in a cross section at the middle of the square side, (c) configuration of the crush zone (Mode Ib) through the wall thickness of a 15° specimen (cross section AA′ in Figure 8.1(a)), (d) micrograph showing microfailures in the crush zone (Mode Ib) for a dynamically loaded 15° square frustum (sp. 33; see Table 8.1) in a cross section at the middle of the square side.

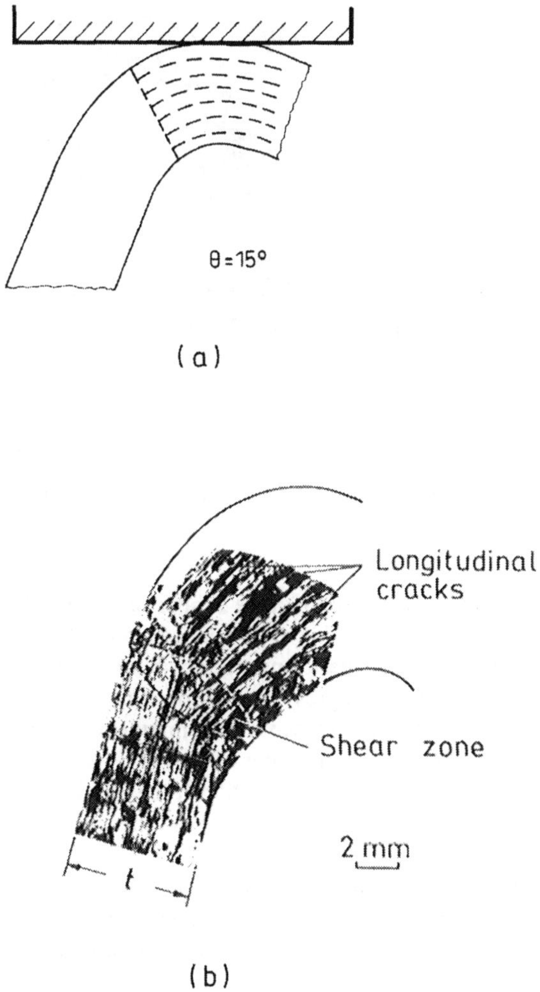

Figure 8.4. (a) Configuration of the crush-zone in the corner of a 15° specimen (cross-section BB′ in Figure 8.1(a)), (b) micrograph showing microfailures in the crush-zone (Mode Ia) for a statically loaded 15° square frustum (sp. 13; see Table 8.1) in a cross section at the corner of the frustum.

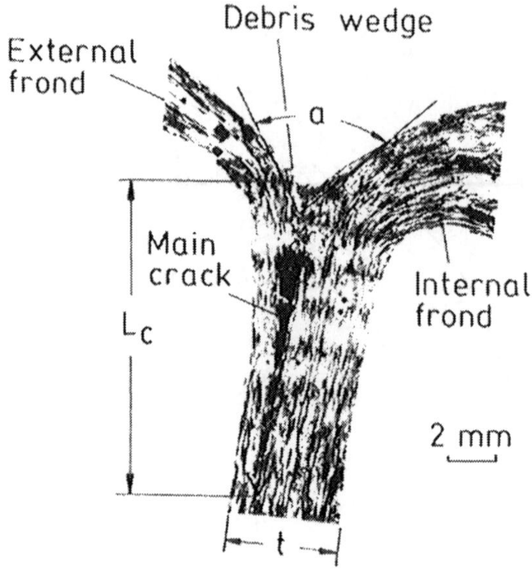

Figure 8.5. Micrograph showing microfailures in the crush zone (Mode Ia) for a dynamically loaded 5° square frustum (sp. 23; see Table 8.1) in a cross section at the middle of the square side.

and buckle when the load, or the length of the lamina bundle, reaches a critical value. At this stage, a triangular debris wedge of pulverised material starts to form, see Figures 8.2 and 8.3(a) and (b); its formation may be attributed to the friction between the bent bundles and the platen of the press or the drop mass. As loading proceeds further, resulting in crushing with the subsequent formation of the internal and external fronds, normal stresses develop on the sides of the debris wedge, followed by shear stresses along the same sides, due to the friction at the interface between the wedge and the fronds. Note, also, that additional normal and shear stresses develop at the interface between the steel press platen or the drop mass and the deforming shell, as the formed fronds slide along this interface. The crushing load in the post-crushing region is characterised by oscillations about a mean post-crushing load, $\overline{P}$; these oscillations start as soon as the formation of the debris wedge is completed.

Regarding the microfracture mechanism of the square frusta subjected to axial loading, as far as the Mode Ia of collapse is concerned, the experimental observations made are similar to those obtained during the axial collapse of circular frusta; see also the remarks reported in Chapter 7. Since the circumference of the shell increases as the crushing of the frusta progresses, it is evident that the size of the wedge increases during crushing. With increasing semi-apical angle of the frustum, the position of the intrawall crack moves towards the outside edge of the shell wall, increasing in this manner the thickness of the inner frond and, simultaneously, resulting in a positioning of the annular wedge mainly above it. On the contrary, the crack length decreases with increasing semi-apical angle; compare Figures 8.2(b), (d) and 8.3(b). From measurements of the specimens tested, the wedge angle, α is about 80°, 70° and 60° for 5°, 10°, and 15° frusta, respectively. The main intrawall crack length diminishes from the centre of the square side towards the square corners, where the typical crush zone disappears; see Figure 8.4 and the similar remarks made in Section 6.3.2 of Chapter 6, regarding the collapse of square tubes. The maximum value of the crack height, L_c, which is attained at the middle of each side of the square cross-section, is almost the same as the corresponding one observed in the case of circular frusta of the same semi-apical angle and loaded under the same conditions, see Section 7.3.2 of Chapter 7. The combined mechanism, developed for the various regions of a compressed square frustum during the crushing process, is shown schematically in Figure 8.1(a).

The Mode Ib of collapse was mainly observed for dynamically loaded specimens, and for higher semi-apical angles. Progressive collapse developed by successive shearing of the region near the narrow end of the shell, accompanied by a number of delaminations and longitudinal cracks, whilst the tube wall inverses inwards; see Figures 8.1(c) and 8.3(c) and (d).

Based on the above description, regarding the micromechanism occurring during the axial collapse of the square frusta, as far as the collapse Mode I is concerned, the total energy dissipated may be attributed to:

- The energy absorbed during the propagation of the various cracks, the development of delaminations and the axial splitting of the shell
- The energy absorbed, because of microfractures and/or flexural damage and the bending of the plies of the fronds
- The energy absorbed due to the development of frictional stresses at the various regions, as outlined above

LONGITUDINAL CORNER CRACKING (MODE II)

The longitudinal corner cracking mode is characterised by the formation of a crack at the corner of the frustum, which propagates towards the narrow end of the specimen, see Figure 8.1(d). A single corner started to crack, usually followed by the cracking of the diametrically opposite corner, whilst subsequently the other two corners started to collapse, providing that the crushing was allowed to continue. The load/displacement curves for the Mode II of collapse do not display as much energy absorbing potential as the Mode I curves; compare Figures 8.6(a) and (b). While the load was developed abruptly, and sometimes there was a distinctive P_{max}, most often the collapse started as the load reached its maximum level and then the extent of progressive crushing was terminated at the early stages of shell shortening, see Figure 8.6(b), exhibiting, therefore, rather low energy absorption capability.

MID-LENGTH COLLAPSE MODE (MODE III)

Specimens followed this collapse mode exhibited extensive brittle failure with a circumferential fracture of the material, see Figure 8.1(e). Fracture started at a distance from the loaded end of the specimens, approximately equal to the mid-height of

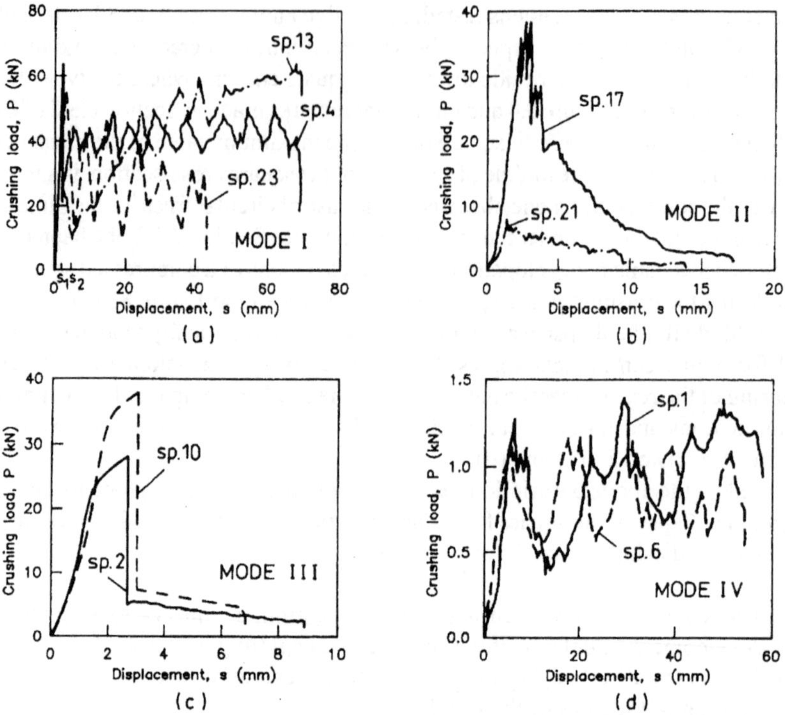

Figure 8.6. Load/displacement curves for the various collapse modes (see Table 8.1).

the shell, and involved catastrophic failure by cracking and separation of the shell into irregular shapes, probably due to local severe shear straining of the wall of the shell. This failure mode is similar to the Euler column-buckling of very thin metallic and PVC tubes subjected to axial loading [8]. The load/displacement curves, corresponding to the mid-length collapse mode, showed a typical pre-crushing region, but the initial elastic response was followed by a very sharp drop in load and poor post-crushing characteristics; see Figure 8.6(c).

PROGRESSIVE FOLDING (MODE IV)

The progressive folding mode of collapse is characterised by the formation of a series of folds or fracture hinges as the frustum crushed, see Figure 8.1(f). The hinges seem to form in regions where the relative maximum deflections occurred during the elastic deformation of the shell. Fairly regular sized plates were formed between the hinges and remained relatively undamaged. This mode of collapse is similar to the collapse of metal and thermoplastic tubes and frusta, where plastic hinges are formed and the specimen proceeds to fold as the crushing proceeds [8]. The load/displacement curves for the Mode IV showed large fluctuations and the average collapse load was relatively small, see Figure 8.6(d).

Note that from the four failure modes observed, the end-crushing mode (Mode I) was found to offer the highest energy absorption capability.

As far as the collapse modes of the square frusta under axial loading is concerned, it must be noted that, for the statically loaded ones all above described modes of failure were observed. The characteristic modes of collapse obtained throughout the dynamic tests can be identified and classified as stable and unstable collapse modes. However, at impact, due to the dynamic nature of the phenomenon, unstable modes of collapse lead rapidly to a complete catastrophic failure, preventing in this manner the investigation of the deformation mechanism. Thus, the only information on this topic may be drawn from the change of the corresponding load/displacement curves. On the contrary, throughout the static test series and due to the wide range of geometric combinations that were used, see Table 8.1(a), the effect of the various geometric parameters, i.e. wall thickness, t, semi-apical angle, θ, mean circumference, C, and axial length, L on the collapse modes was studied.

Regardless of thickness, frusta with smaller semi-apical angles (0°–15°) failed following stable modes of collapse. Frusta with large semi-apical angles, i.e. 20°–30°, followed the Mode II of collapse, associated with longitudinal corner cracking. This is a low energy collapse mode and quite inappropriate for energy absorption. Specimens of smaller semi-apical angles did not exhibit the Mode II collapse, however, some collapsed following the Mode III, which is also a low energy collapse mode. These observations are indicative of the existence of a transitional zone, with the frustum semi-apical angle ranging from 15° to 20°, where the collapse mode changes from a stable to an unstable one. Note that circular frusta exhibit similar behaviour, for semi-apical angles greater than 20°; see Section 7.3.2 of Chapter 7. Frusta of large semi-apical angles showed no significant change of the collapse mode with increasing wall thickness of the shell. Frusta of 5°–15° semi-apical angles, with wall thickness of 1.5 mm or less, statically loaded, followed the progressive folding Mode IV of collapse. Frusta of the same semi-apical angles, with wall thickness greater than 2.5 mm, exhibited the Mode I of collapse when loaded at elevated strain-rates. Frusta of semi-apical angles of 5° or 10°, with wall thickness 1.5 mm $< t <$ 2.5 mm, exhibited the Mode III of collapse when subjected to static loading. On the contrary, the unstable mode of collapse Mode II was observed in the case of dynamically loaded frusta of 15° semi-apical angle and 2.6mm wall thickness.

The experimentally obtained deformation modes of all specimens tested are classified in respect to the geometry factors, wall thickness/mean circumference, t/C, and axial length/mean circumference, L/C, and are presented in Figure 8.7. Distinct regions, characteristic for the various deformation modes developed, and the transition boundaries from stable to unstable modes of collapse are indicated, providing, therefore, useful information about the collapse of the square frusta and their behaviour as energy absorbers. Shell instability occurs for values of the geometric factors t/C and L/C lower than critical ones, see Figure 8.7; these critical values, which are almost identical for static and dynamic loading, are about 0.001 for t/C and 0.4 for L/C. Note that the critical value of L/C is almost equal to the related one for circular frusta, subjected to similar loading conditions, whilst the value of t/C is much higher for the square frusta, as compared to the circular ones; see also Section 7.3.2 of Chapter 7.

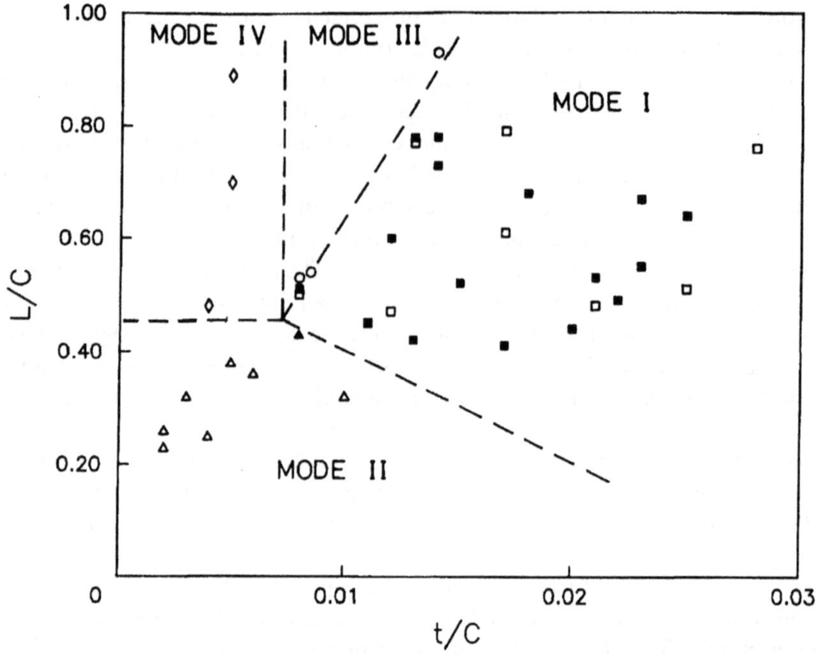

Figure 8.7. Classification chart showing the areas of collapse modes and transition boundaries from one mode to another for composite square frusta (□ - Mode I-static; ■ - Mode I - Impact; △ - Mode II-static; ▲ - Mode II-impact; ○ - Mode III-static; ◇ - Mode IV-static).

As far as the Mode I of collapse is concerned, the initial axial length of the shell, L seems not to affect the height of the main central crack, L_c and the mechanical response of the shell due to failure, see Table 8.1. As outlined above, the tip of the main intrawall crack is shifted towards the outer edge of the shell wall, as the frusta semi-apical angle increases, whilst the size of the wedge, as well as the height of the main crack, L_c, reduce, owing to the transition from Mode Ia to Mode Ib, see Figures 8.1(b), (c) and 8.3.

Dynamic loading greatly affects the microfracture mechanism, the size of the debris wedge and the main crack developed, because of the wedge propagation, are smaller, whilst the tip of the crack is closer to the outer wall surface for impacted shells, as compared to statically loaded ones; compare Figures 8.3 (b) and 8.5. Therefore, transition from Mode Ia to Mode Ib, see Figures 8.1(b) and (c), occurs earlier with increasing strain-rate; compare Tables 8.1(a) and (b). This is in agreement with the remarks reported during the loading of circular frusta at elevated strain rates, see Reference [82] and also Section 7.3.2 of Chapter 7. Note that the characteristics of the crush zone observed at the four corners of the frustum are similar to those obtained for the crush zone of Mode Ib, created by successive shearing. For the dynamically loaded specimens, it was observed that some regions, near the crush zone and near the four corners, turn opaque white, probably due to the release of very strong shock waves that were caused by the violent crash between the drop-mass and the shell, causing instantaneous propagation of delamination cracks, accompanied with abrupt drops of the crushing load. This phenomenon is more profound for square frusta with large semi-apical angles.

8.3.3 Energy Absorbing Characteristics

In Figure 8.6 the crushing load/shell shortening curves for the various collapse modes of statically and dynamically loaded shells are presented. In general, the end-crushing mode Mode I seems to offer the highest energy absorption capability in relation to the other three collapse modes.

Initially the shell behaves elastically until the initial maximum value of the load, P_{max} is reached. It is greatly affected by the shell geometry, i.e. the wall thickness (number of layers), the semi-apical angle and the mean circumference of the shells, the material characteristics and the corners rigidity, see Table 8.1. For constant semi-apical angle the peak load, P_{max} increases with increasing wall thickness, whilst for the same number of layers, i.e. for constant wall thickness, decreases with increasing semi-apical angle, but increases with increasing mean circumference. Note that the above mentioned observations were made on frusta with the same triggering mechanism, i.e. tapered top of the frustum properly polished and free of microdefects.

As deformation progresses, the shape of the load/displacement curve depends on the mode of collapse; for thin-walled composite shells, it was observed that the fracture behaviour of the shell appears to affect the loading stability, as well as the magnitude of the crush load and the energy absorption during the crushing process. It may be assumed that, at any instant, the crush load must be supported by more than one structural elements of the shell and, furthermore, a specific element contributes more to supporting the load; see also Reference [81].

The mean post-crushing load, $\overline{P}$ and the energy absorbed, W increase considerably with increasing thickness or number of layers of wrapped glass mat. However, due to the fact that the radius of curvature of the internal frond, r_i increases with increasing semi-apical angle, whilst the external frond is constrained to deform through a smaller radius r_o, see Figures 8.2 and 8.3, the total energy dissipated depends directly upon the semi-apical angle; see Table 8.1 and the similar remarks concerning the axial loading of circular frusta reported in Section 7.3.3 of Chapter 7. Therefore, for a given slenderness ratio, t/C, the specific energy decreases as the semi-apical angle increases, see Table 8.1. Note, also, that there is no effect on the increase of specific energy for specimens with number of layers greater than 3; see Table 8.1 for details. This observation is in agreement with those reported in References [83] and [82] concerning circular frusta statically and dynamically loaded, respectively. For a given angle, the mean post-crushing load, $\overline{P}$ increases with increasing slenderness ratio, t/C, whilst the mean load tends to increase as the semi-apical angle increases with the slenderness ratio kept constant; see Table 8.1.

For dynamically loaded frusta, the effect of the strain-rate on the specific energy and the mean post-crushing load seems to be almost negligible, see Table 8.1(b). For specimens with three or more layers, peaks and valleys of the load occur for approximately the same displacement, indicating, therefore, that the specific energy remains practically constant. For relatively thick shells (4–6 layers), the valleys of the load approach zero, probably due to the fact that a secondary deformation mechanism developes after the intense cracking and delamination.

The load/displacement curves have several similar features with the corresponding ones of collapsed square tubes made of the same material. However, it must be noted that a distinct difference, pertaining to the post-crushing region of the curves between square tubes and frusta (with large semi-apical angle) was observed. The crushing load tended to increase as crushing progressed, due to the increase of the effective cross-sectional area of the frusta being crushed; compare Figure 8.6(a) with Figure 6.4 of Chapter 6. The duration of the crushing phenomena depends upon the shell geometry, material properties and the drop mass and height. However, from the shape of the dynamically obtained curves in the post-crushing region, it renders difficulties, as far as the possible fracture mechanism occurred, and the development and propagation of microcracks during dynamic loading is concerned. As reported in Chapter 5, two different fracture mechanisms may be proposed, based on observations related to crack propagation in composite materials, neglecting friction forces and bending, as well as microscopic observations concerning the collapse mode of shells.

For the frusta subjected to static and dynamic loading, it is evident that with increasing the geometrical factor t/C, i.e. by increasing wall thickness or decreasing mean circumference, the specific energy absorbed increases.

8.3.4 Failure Analysis

STATIC AXIAL COLLAPSE

Various crashworthy phenomena, pertaining to the axial collapse of composite multi-layered shells, are associated with the distribution of the absorbed energy during the crushing process. In the proposed theoretical approach, see Figure 8.2(a), the following crushing phenomena were encountered: friction between the annular wedge and the fronds and between the fronds and the platen of the press; fronds bending; crack propagation; axial splitting. The theoretical model proposed in Section 6.3.4 of Chapter 6 and Section 7.3.4 of Chapter 7, for the analysis of composite square tubes and circular frusta, respectively, subjected to static axial loading, was modified and used to analyse the collapse mechanism and to estimate the related energy absorbed during the axial crushing of the square frusta.

During the elastic deformation of the shell, the load rises at a steady rate to a peak value, P_{max}, see Figure 8.6. At this stage, cracks of height L_c and length $L_c/\cos\theta$ form at the four corners of the shell and propagate downwards along the frustum axis, splitting the shell wall, see Figure 8.1(b); they are accompanied by the development of a circumferential main intrawall crack at the top end of the shell, whilst the related shell shortening is s_l, see Figure 8.6(a). It is assumed that the crack length distribution along the circumference of the frustum follows an elliptical path, as shown in Figure 8.1(a). The maximum crack length, $L_c/\cos\theta$ is attained in the middle of each side of the square cross section and it is almost equal to the corresponding one, observed in the case of the equivalent circular frusta loaded under the same conditions, see Section 7.3.4 of Chapter 7. It must be noted that the position of the occurrence of the main intrawall crack moves away from the shell wall axis towards the outside edge of the shell wall with increasing semi-apical angle of the frustum, θ, whilst the crack height, L_c decreases; see Figure 8.2. The frustum side width at the crack tip, b_c, related to the position of the crack initiation, as measured experimentally, is

$$b_c = (12 \cdot t \cdot \theta/\pi) + b_2 - t \tag{8.1}$$

where, following the notation, t is the shell wall thickness and b_2 the shell outside top width.

Therefore, the associated absorbed energy, which equals the external work, as obtained by measuring the area under the load/displacement curve in the elastic regime in Figure 8.6(a), is

$$\begin{aligned} W_{Lc} &= 2 \cdot [\pi \cdot (L_c/\cos\theta) \cdot (b_c/2)] \cdot R_{\mathrm{ad}} + n \cdot [(t/2) - 6 \cdot \theta \cdot (t/\pi)] \cdot G \\ &\cdot (L_c/\cos\theta) = \int_0^{s_1} P ds = \frac{1}{2} P_{\max} s_1 \end{aligned} \tag{8.2}$$

where, R_{ad} is the fracture energy required to fracture a unit area of the adhesive at the interface between two adjacent layers, n the number of splits and G the fracture toughness.

The energy required for the deformation mechanism regarding the history of the formation of the crush zone, see Section 7.3.4 of Chapter 7, equals the external work absorbed by the deforming shell in this regime, i.e.

$$\begin{aligned} W_{\mathrm{tr}} &= \left[\int_0^{\alpha_1+\theta} \sigma_o \cdot l_{s1} \cdot (l_{s1}/2) d\alpha + \int_0^{\alpha_2-\theta} \sigma_o \cdot l_{s2} \cdot (l_{s2}/2) d\alpha \right] \\ &\cdot 4 \cdot (b_c + s \cdot \tan\theta) = \int_{s_1}^{s_2} P ds \end{aligned} \tag{8.3}$$

where, according to the notation, σ_o is the normal stress applied by the wedge to fronds, l_{s1} the side length of the wedge inscribed to the external bent frond, l_{s2} the side length of the wedge inscribed to the internal bent frond, α_1 the angle formed by the height and the external side of the wedge, α_2 the angle formed by the height and the internal side of the wedge, see Figures 8.1(a) and 8.2(a); s_2 is the related shell shortening corresponding to the completion of the wedge formation, see Figure 8.6(a).

Since the intrawall crack propagates at a constant speed equal to the speed of the crosshead of the press, it may be assumed that the crack height, L_c remains constant.

Also the height of the wedge, h, equals $0.7 \cdot (t/\cos\theta)$, as observed experimentally, see Figures 8.2 and 8.3. Therefore, taking into account the failure mechanism outlined above, the total dissipated energy for a crush distance, s can be estimated as follows:

- Energy dissipated due to friction between the annular wedge and fronds and between fronds and platen

$$W_i = (\mu_{s1} \cdot (P_1 + P_2) + \mu_{s2} \cdot (P_3 + P_4)) \cdot 4 \cdot b \cdot (s - s_2) \tag{8.4}$$

where, P_1 and P_2 are the normal forces per unit length applied by the platen to the internal and external fronds, respectively, which result from static equilibrium, see also Section 7.3.4 of Chapter 7; P_3 and P_4 are the normal forces per unit length applied to the internal and external sides of the wedge, respectively, μ_{s1} is the coefficient of friction between frond and platen, μ_{s2} the coefficient of friction between the wedge and the fronds and $b = b_c + s \cdot \tan\theta$. Note that

$$P_3 = \sigma_o \cdot l_{s1} \tag{8.5}$$

$$P_4 = \sigma_o \cdot l_{s2} \tag{8.6}$$

and

$$\sigma_o = k \cdot \sigma_\theta \tag{8.7}$$

therefore,

$$\begin{aligned} W_i = \{ & \mu_{s1} \cdot [(W_T / 4 \cdot b \cdot s) - (0.7 \cdot t \cdot k \cdot \sigma_\theta / \cos\theta) \\ & \cdot (\tan a_1 + \tan\alpha_2 + 2 \cdot \mu_{s2})] + \mu_{s2} \cdot (0.7 \cdot t \cdot k \cdot \sigma_\theta / \cos\theta) \\ & \cdot (1/\cos\alpha_1 + 1/\cos\alpha_2)\} \cdot 4 \cdot b \cdot (s - s_2) \end{aligned} \tag{8.8}$$

where, W_T is the total energy dissipated for the deformation of the shell, k is a constant and σ_o the tensile fracture stress of the composite material.

- Energy dissipated due to fronds bending

$$\begin{aligned} W_{ii} = \{ & \int_0^{\alpha_1+\theta} P_3 \cdot (l_{s1}/2) d\alpha + \int_0^{\alpha_2-\theta} P_4 \cdot (l_{s2}/2) d\alpha + \int_{s_2}^{s} [P_3 \cdot (\alpha_1 + \theta) + P_4 \\ & \cdot (\alpha_2 - \theta)] ds\} \cdot 4 \cdot b = (2.8 \cdot t \cdot k \cdot \sigma_\theta \cdot b / \cos\theta) \\ & \cdot \{[(\alpha_1 + \theta)/\cos\alpha_1)] \cdot [035 \cdot t/(\cos\alpha_1 \cdot \cos\theta) + s - s_2] \\ & + [(\alpha_2 - \theta)/\cos\alpha_2)] \cdot [035 \cdot t/(\cos\alpha_2 \cdot \cos\theta) + s - s_2]\} \end{aligned} \tag{8.9}$$

- Energy dissipated due to crack propagation

$$W_{iii} = R_{ad} \cdot [(s - s_1) \cdot 4 \cdot b + \pi \cdot (L_c / \cos\theta) \cdot b_c] \quad (8.10)$$

- Energy dissipated due to axial splitting

$$W_{iv} = 4 \cdot [(t/2) - 6 \cdot \theta \cdot (t/\pi)] \cdot G \cdot s \quad (8.11)$$

From Equations (8.9–8.11) the total energy dissipated for the deformation of the shell is obtained as

$$\begin{aligned} W_T = {} & [1/(1 - \mu_{s1} + \mu_{s1} \cdot s_2 / s)] \cdot \{[2.8 \cdot t \cdot k \cdot \sigma_\theta \cdot (b/\cos\theta)/\cos\theta] \\ & \cdot [(s - s_2) \cdot [(\mu_{s2} \cdot (1/\cos\alpha_1 + 1/\cos\alpha_2) - \mu_{s1} \\ & \cdot (\tan\alpha_1 + \tan\alpha_2 + 2 \cdot \mu_{s2})] + [(\alpha_1 + \theta)/\cos\alpha_1] \\ & \cdot [0.4 \cdot t/(\cos\alpha_1 \cdot \cos\theta) + s - s_2] + [(\alpha_2 - \theta)/\cos\alpha_2] \\ & \cdot [0.35 \cdot t/(\cos\alpha_2 \cdot \cos\theta) + s - s_2]] + R_{ad} \cdot [(s - s_1) \\ & \cdot 4 \cdot b + \pi \cdot (L_c / \cos\theta) \cdot b_c] + 4 \cdot [(t/2) - 6 \cdot \theta \cdot (t/\pi)] \cdot G \cdot s\} \end{aligned} \quad (8.12)$$

whilst the total normal force applied by the platen to the shell can be calculated as

$$P = W_T / s \quad (8.13)$$

EFFECT OF STRAIN-RATE

As outlined in Section 8.3.2, the microfracture mechanism for the progressive collapse of axially dynamically loaded square frusta is, in general, similar to that obtained during the axial static collapse. The only differences encountered are dealing with the shape and the position of the wedge and the microcracking development; compare Figures 8.3(b) and 8.5. Therefore, the above proposed theoretical model can be used for the prediction of the crushing loads and the energy absorbed in the case of circular frusta subjected to axial dynamic loading, with the assumption that the tensile fracture stress and the fracture toughness of the material remain constant, whilst these parameters can be a function of the crush-speed.

As predicted both theoretically and experimentally for other shell geometries, the major part of the absorbed energy during the static axial compression of a shell is dissipated as frictional work in the crushed material or at the interface between the material and the tool and, this is estimated to be about 50% or more of the total work done; see Sections 5.3.4, 6.3.4 and 7.3.4 of Chapters 5, 6 and 7, respectively. In dynamic collapse, attention is directed towards the influence of strain-rate on the frictional work absorbed during the impact, taking into account all structural and material parameters which may contribute to it, i.e. the fibre and matrix material, the fibre diameter and orientation in the laminate, the fibre volume content, as well as the conditions at the interfaces.

As mentioned above, two distinct regions, where the development of frictional forces is of great importance, were identified: the fronds/wedge contact region, composed of the same material, and the fronds/platen contact area, composed of different materials. The coefficient of friction in these regions depends on various phenomena, associated with the material flow described in detail in Section 6.3.4 of Chapter 6. These phenomena may determine the increase or the decrease of the friction coefficient; they are greatly affected by the material properties influencing the effectiveness of the collapsed, component as far as its crashworthy capacity is concerned.

The amount of energy dissipated due to fronds bending, W_{ii}, to crack growth, W_{iii} and to axial splits, W_{iv} are estimated from Equations (8.9), (8.10) and (8.11) respectively, whilst the energy dissipated due to friction, W_i is estimated from Equation (8.8) by substituting the static coefficients, μ_{s1} and μ_{s2} with the related dynamic ones, μ_{d1} and μ_{d2}, respectively. Therefore, the total energy absorbed during the dynamic crushing of the shells can be calculated from Equation (8.12), properly modified, as

$$\begin{aligned} W_T = {} & [1/(1-\mu_{d1} + \mu_{d1} \cdot s_2/s)] \cdot \{[2.8 \cdot t \cdot k \cdot \sigma_\theta \cdot (b/\cos\theta)/\cos\theta] \\ & \cdot [(s-s_2) \cdot [(\mu_{d2} \cdot (1/\cos\alpha_1 + 1/\cos\alpha_2) - \mu_{d1} \\ & \cdot (\tan\alpha_1 + \tan\alpha_2 + 2 \cdot \mu_{d2})] + [(\alpha_1 + \theta)/\cos\alpha_1] \\ & \cdot [0.4 \cdot t/(\cos\alpha_1 \cdot \cos\theta) + s - s_2] + [(\alpha_2 - \theta)/\cos\alpha_2] \\ & \cdot [035 \cdot t/(\cos\alpha_2 \cdot \cos\theta) + s - s_2]] + R_{ad} \cdot [(s-s_1) \cdot 4 \\ & \cdot b + \pi \cdot (L_c/\cos\theta) \cdot b_c] + 4 \cdot [(t/2) - 6 \cdot \theta \cdot (t/\pi)] \cdot G \cdot s\} \end{aligned} \tag{8.14}$$

From the stress-strain curve for the composite material A shown in Figure 5.4 of Chapter 5, the tensile fracture stress, σ_θ of the material was estimated. The static friction coefficients, μ_{s1} and μ_{s2} were obtained by employing the curling test [41]. An estimate of the dynamic coefficients of friction at the fronds/platen interface, μ_{d1} and the fronds/wedge interface, μ_{d2} was obtained by comparing the experimental crushing loads in Table 8.1. The fracture toughness, G was estimated from the tension test of notched strips. The interfacial fracture energy, R_{ad} was calculated from Equation (5.1) of Chapter 5, using the experimental results obtained by loading cylindrical tubes up to the maximum load, P_{max} and obtaining the energy absorbed from the related experimental load/displacement curves of loaded shells. However, as mentioned there, the R_{ad} may be also estimated by employing the strip peel test, see Reference [46].

8.3.5 Crashworthy Capability: Concluding Remarks

The effect of the specimen geometry and the loading rate on the energy absorbing capability of thin-walled fibre-reinforced square frusta subjected to axial collapse, was both experimentally and theoretically investigated. The collapse modes at macroscopic scale and the determination of the microfailures during the collapse of the structure tested were also examined and analysed.

Four collapse modes were observed:

- The stable progressive collapse mode, Mode I, associated with large amounts of crush energy, resulting, therefore, in a high crashworthy capacity of the structural component
- The longitudinal corner cracking, Mode II, showing successive formations of cracks at the shell corners
- The mid-length collapse mode, Mode III, featuring a brittle fracture involving catastrophic failure
- A progressive folding mode, Mode IV, with the formation of sharp hinges

Square frusta with large mean circumference were more prone to catastrophic failure. Shell buckling instability is prevented by ensuring that the *t/C* and *L/C* ratios are above a critical value, see Figure 8.7. These values seem to be almost equal for dynamic and static tests, respectively. The *t/C* ratio overestimates the corresponded one of the circular tubes and frusta under axial loading.

As far as the behaviour of square frusta under static and dynamic loading is concerned, the variation of characteristic parameters, i.e. specific energy and mean post-crushing load, is similar to that observed during the axial collapse of circular frusta at elevated strain-rates, i.e. the specific energy decreases with increasing semi-apical angle and the mean-post crushing load increases with increasing wall thickness, see also Chapter 7. Note, also, that under dynamic loading, and for specimens with number of layers greater than 3, no significant effect on the increase of the specific energy was observed.

From the experimental results, it is indicated that static tests, for the same shell geometries, develop higher values of specific energy than these obtained in dynamic loading, probably due to the higher values of the static friction coefficients, μ_{s1} and μ_{s2} between the wedge and the fronds and between the platen surface and the fronds (about 10–15% of the related dynamic ones); this increase in the crashworthy ability of the shell was about 5–15%. This is in agreement with the results reported in Chapter 7, concerning the crashworthy characteristics of circular frusta at elevated strain-rates. As observed, the increase of the specific energy, for the circular frusta statically loaded, is greater than for the corresponded square ones for all slenderness ratios, whilst for the dynamically loaded ones the specific energy seems to be slightly influenced by the geometrical differences of the shells.

The mean post-crushing load, $\overline{P}$, the total energy absorbed, W and the specific energy, W_s are well predicted theoretically by the proposed analysis, within $\pm 12\%$, see Table 8.1. According to the proposed theoretical analysis, the distribution of the dissipated energy of the crush shell due to the four main energy sources was estimated as:

- Energy due to friction between the annular wedge and the fronds and between the fronds and the platen, about 45%, 47.5% and 50% of the total energy dissipated for the 5°, 10° and 15° square frusta, respectively
- Energy due to fronds bending, about 48%, 46% and 44%, respectively
- Energy due to crack propagation, about 6%, 5.5% and 5%, respectively

- Energy due to axial splitting at the four corners of the shell, about 1% for all specimens tested

It is evident that the contribution of the frictional conditions between wedge/fronds and fronds/platen to the energy absorbing capability is more significant than the other ones. As discussed above, it mainly depends upon the friction coefficients, μ_{s1}, μ_{d1} and μ_{s2}, μ_{d2}, which are affected by the surface conditions at the interfaces between composite material/platen or drop mass and composite material/debris wedge, respectively.

Classifying the geometries tested, from the energy absorbing capacity point of view, the geometry of a 5° square frustum seems to be the more efficient for both testing conditions, see Table 8.1. This is in agreement with those observed for circular frusta axially loaded and reported in Chapter 7.

Regarding the microfracture mechanism of square frusta subjected to axial loading, as far as the Mode Ia of collapse is concerned, the experimental observations lead to the conclusion that, the microfracture mechanism may be determined as a combination of the main characteristics observed for square tubes and circular frusta. In general, this mechanism is similar for statically and dynamically loaded shells, the only differences encountered being associated with the shape of the wedge and the microcracking development. The transition from Mode Ia to Mode Ib occurs earlier for the dynamically loaded frusta.

CHAPTER 9

AUTOMOTIVE SECTIONS

9.1 NOTATION

C = mean circumference of shell
D = height of the beam cross section
E = Young's modulus
G = fracture toughness
G^* = shear modulus
g = acceleration of gravity
I = moment of inertia
k = constant
L = axial length of tube
L_c = length of central crack
l_s = side length of pulverised wedge
M = bending moment
M_{max} = maximum bending moment
n = number of axial splits
P = current crushing load
P_1, P_2 = normal force per unit length
$\overline{P}$ = mean post-crushing load
P_{max} = peak load
R_{ad} = fracture energy per unit area of layers
S = shear strength
s = displacement, shell shortening, crush length
t = wall thickness
v = crush-speed
W = energy absorbed
W_s = specific energy
W_T = total energy dissipated
W_{tr} = energy required for the crush zone formation

w = deflection
X_c, X_t = compressive and tensile strength of lamina in longitudinal direction
Y_c, Y_t = compressive and tensile strength of lamina in transverse direction
α = angle of wedge
γ = acceleration
ε = strain
θ = angle of rotation
μ_s = static friction coefficient
μ_d = dynamic friction coefficient
ν = Poisson's ratio
σ_θ = tensile fracture stress
σ_o = normal stress
φ (=α/2) = semi-angle of wedge

9.2 GENERAL

An automotive frame rail of hourglass cross section, made of a glass fibre/vinylester composite, was designed for use in the apron construction of the car in order to obtain a high degree of crashworthiness at this location of the car body. This Chapter deals with the crashworthy behaviour of this structural component in axial compression at various strain-rates (head-on collision) and in bending (oblique collision), see References [77, 78, 112]. The modes of collapse at macroscopic scale, the microscopic fracture patterns and the energy absorbing capability of such rail beams were examined and discussed. A theoretical analysis of the collapse mechanism of the components, tested under axial loading, is proposed leading to a good approximation of the energy absorbed during crushing. Moreover, the ultimate strength of cantilevered specimens, subjected to bending, was also theoretically predicted and found to be in good agreement with the experimental results.

9.3 AXIAL COLLAPSE: STATIC AND DYNAMIC

9.3.1 Experimental

The material used, designated as composite material B, was a commercial glass fibre and vinylester composite. The tube wall consisted of nine plies with a total thickness of 3.3 mm for the axial collapsed specimens. Starting from the exterior of the shell the plies were laid-up in the sequence $[(90/0/2R_c)/(2R_c/0/90)/R_{c.75}]$, where the 0° direction coincidet with the axis of the tube, R_c denotes random chopped strand mat plies and $R_{c.75}$ represents a similar ply but thinner, providing, in this manner, with a composite material of 33.9% per volume fibre content and 1.55 g/cm^3 density. Fur-

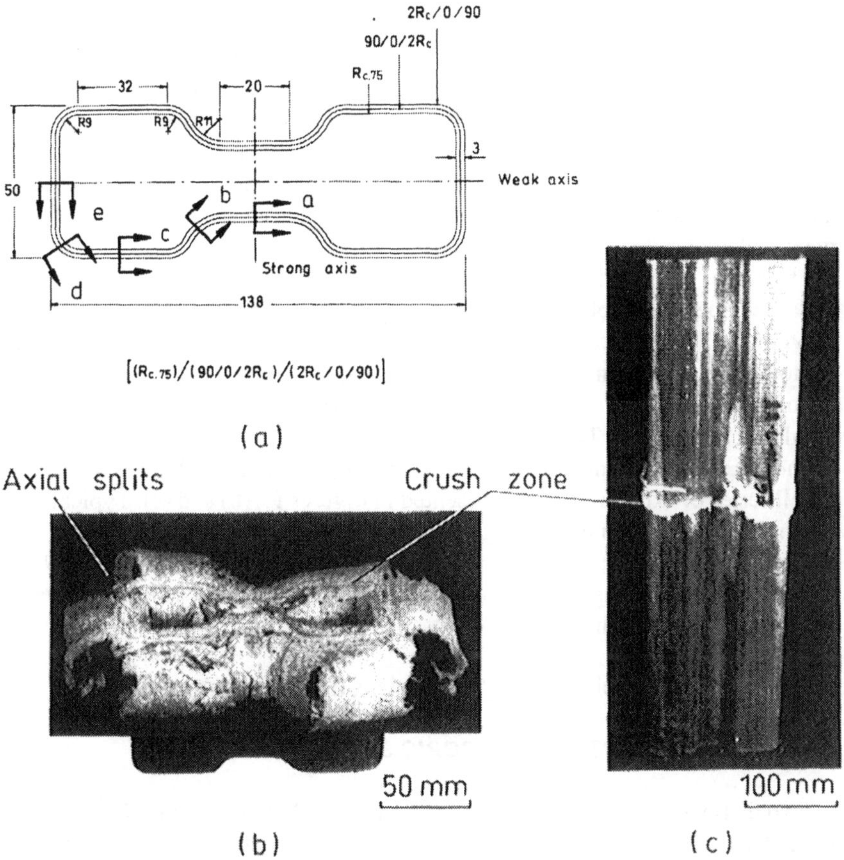

Figure 9.1. (a) cross section of the composite hourglass rail shell; (b) macroscopic view of Mode I of collapse (sp. 5; see Table 9.1); (c) macroscopic view of Mode II of collapse (sp. 15; see Table 9.1).

ther static axial collapse tests, with tubes made of a material, designated as material *B1*, similar to *B*, but with 3 extra plies laid-up in the sequence $[(90/0/2R_c)/(\pm45)/(2R_c/0/90)/R_{c.75}]$ from the outside towards the inside surface and with a total wall thickness of 4.3 mm, were also carried out. Note that, due to the fabrication procedure, both specimens were slightly thicker in their curved parts, see b in Figure 9.1(a), because of excessive amount of resin, however, the thickness deviation was very small and, therefore, a constant thickness was considered throughout the present work. Details about the fabrication of the composite shell and the properties of both materials, as well as about the lay-up, are presented in Section 5.3.1 of Chapter 5. A stress-strain curve of the material as obtained from quasi-static tension test is illustrated in Figure 5.4 of Chapter 5 (material B).

The static axial collapse was carried out between the parallel steel platens of a SMG hydraulic press at a crosshead spead of 10 mm/min or a compression strain rate of 10^{-3} sec^{-1}. The corresponding dynamic tests were performed by direct impact on a drop-hammer equipped with a 75 kg falling mass. The drop-height was 2.8 m, resulting in an initial impact velocity of 7 m/s or a compression strain rate of 44 sec^{-1}. The experimental set-up and measuring devices used throughout these tests are described in detail in Section 5.3.1 of Chapter 5. Load/tube shortening (displacement) curves during the crushing process were automatically measured and recorded for both types of loading.

The values of the initial peak load, P_{max} and the energy absorbed, W for the axially collapsed specimens, obtained by measuring the area under the load/ displacement curve, as well as the mean crushing load, $\overline{P}$ (defined as the ratio of energy absorbed to the total shell shortening) and the specific energy, W_s (equal to the energy absorbed per unit mass crushed, calculated as the crushed volume times the density of the material) are tabulated in Table 9.1.

Photographs, showing characteristic terminal views of the deformed specimens for all series of experiments, are presented in Figures 9.1(b) and (c). Typical micrographs of the crush-zone, showing the main microfailures, as obtained using a Unimet metallographic optical microscope, are also presented in Figures 9.2 and 9.3. For details about specimens preparation and polishing for the microscopic observations, see Section 5.3.1 of Chapter 5.

9.3.2 Failure Mechanisms: Experimental Observations

COLLAPSE MODES AT MACROSCOPIC SCALE

Two distinct modes of collapse, classified as Mode I and Mode III, respectively, were observed throughout the axial static and dynamic tests. Short specimens, up to a certain length, followed the failure Mode I (progressive collapse), whilst relatively long specimens are characterised by the failure Mode III failure (column-buckling); see the relevant remarks in Chapters 6, 7, and 8.

Mode I of failure, similar to a "mushrooming" failure, is characterised by progressive collapse through the formation of continuous fronds, which spread outwards and inwards, see Figure 9.1(b) and Chapters 5–8. As deformation proceeds further, the externally formed fronds curl downwards with the simultaneous development of a number of axial splits around the circumference of the shell and in positions where there is a distinct change of curvature due to stress concentration, see Figure 9.1(b). It must be noted that the formation of these splits was more distinct around the corners of the shell. Axial tears were not apparent in the internal fronds, which were more continuous than their external counterparts. At the early stage of loading, the shell initially behaved elastically and as soon as the load reached a peak value, depending on shell geometry, material characteristics and corners rigidity, cracks formed at each of the four corners and propagated downwards along the tube axis; they are associated with the formation of a central intrawall crack at the end of the shell adjacent to the loading area, see Figures 9.2(e) and 9.3(e). The post-crushing regime is charac-

Table 9.1: Crushing characteristics of axially loaded hourglass cross-sectioned shells

(a) Static

Sp. No	Mater.	Thick-ness, t (mm)	Axial length, L (mm)	Crush length, s (mm)	Collapse Mode	Crushing load, P (kN) Initial peak, P_{max} Exper.	Mean post-crushing, $\bar{P}$ Exper.	Mean post-crushing, $\bar{P}$ Theor.	Total energy absorbed, W_T(kJ) Exper.	Total energy absorbed, W_T(kJ) Theor.	Specific energy, Ws (kJ/kg) Exper.	Specific energy, Ws (kJ/kg) Theor.
1	B*	3.3	25.4	17.8	Ia	196.5	115.3	116.7	2.046	2.077	54.1	54.9
2	B1*	4.3	25.4	18.0	Ia	250.6	132.6	141.9	2.386	2.554	47.8	51.2
3	B	3.3	50.8	32.5	Ia	183.7	119.5	116.7	3.882	3.783	56.1	54.9
4	B1	4.3	50.8	34.3	Ia	221.5	124.5	141.9	4.262	4.857	44.9	51.2
5	B	3.3	76.2	48.5	Ia	173.5	132.6	116.7	6.431	5.660	62.2	54.9
6	B1	4.3	76.2	50.0	Ia	195.6	141.0	141.9	7.052	7.095	50.8	51.2
7	B	3.3	101.6	65.0	Ia	191.4	118.7	116.7	7.715	7.586	55.8	54.9
8	B1	4.3	101.6	65.0	Ia	288.9	120.6	141.9	7.838	9.224	43.5	51.2
9	B	3.3	152.4	64.5	Ia	195.4	117.7	116.7	7.592	7.527	55.2	54.9
10	B1	4.3	152.4	65.0	Ia	249.9	144.0	141.9	9.357	9.224	51.9	51.2
11	B	3.3	203.2	64.3	Ia	193.2	129.0	116.7	8.289	7.504	60.6	54.9
12	B1	4.3	203.2	65.3	Ia	212.9	140.5	141.9	9.166	9.226	50.6	51.2
13	B	3.3	304.8	64.2	Ia	188.1	109.6	116.7	7.030	7.492	51.4	54.9
14	B1	4.3	304.8	63.8	III	232.9	-	-	2.345	-	-	-
15	B	3.3	508.0	24.3	III	183.2	-	-	0.963	-	-	-
16	B1	4.3	508.0	62.5	III	239.9	-	-	6.633	-	-	-

(b) Dynamic

Sp. No	Mater.	Thick-ness, t (mm)	Axial length, L (mm)	Crush speed, v (m/s)	Crush length, s (mm)	Collapse Mode	Crushing load, P (kN) Initial peak, P_{max} Exper.	Mean post-crushing, $\bar{P}$ Exper.	Mean post-crushing, $\bar{P}$ Theor.	Total energy absorbed, W_T(kJ) Exper.	Total energy absorbed, W_T(kJ) Theor.	Specific energy, Ws (kJ/kg) Exper.	Specific energy, Ws (kJ/kg) Theor.
17	B	3.3	50.8	8.1	11.0	Ia	251.6	153.6	145.5	1.690	1.600	69.7	66.2
18	B	3.3	76.2	8.1	12.9	Ia	212.8	141.2	145.5	1.822	1.877	66.3	66.2
19	B	3.3	101.6	8.1	11.3	Ia	281.0	151.7	145.5	1.714	1.644	68.4	66.2
20	B	3.3	152.4	8.1	25.9	III	255.2	-	-	1.724	-	-	-

* See Section 9.3.1 for materials lay-up

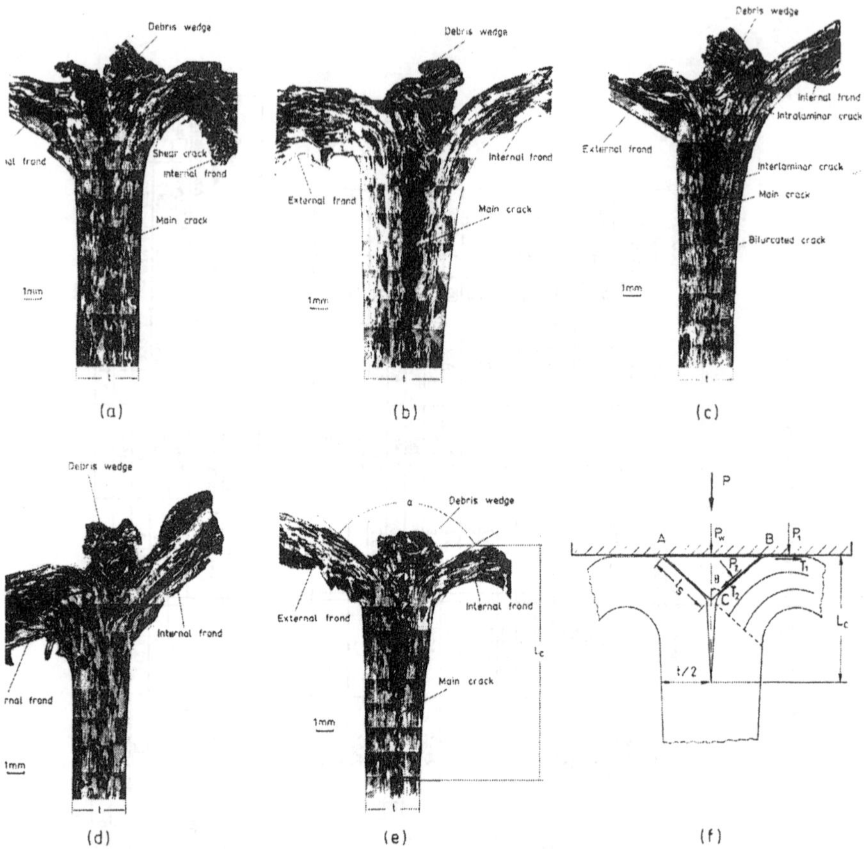

Figure 9.2. Micrographs showing microfailures in the crush-zone for statically loaded hourglass rail shell (sp. 5; see Table 9.1) for positions of Figure 9.1(a): (a) a, (b) b, (c) c, (d) d, (e) e, (f) configuration of the crush zone of (e).

terised by the formation of two equal lamina bundles, bent inwards and outwards, due to the flexular damage which occurs at a distance from the contact surface equal to the wall thickness; they withstand the applied load and buckle when the load, or the length of the lamina bundle, reaches a critical value. At this stage, a triangular debris wedge of pulverised material starts to form, see Figures 9.2(e) and 9.3(e); its formation may be attributed to the friction between the bent bundles and the platen of the press or the drop mass. It must be noted that the wedge size remained constant throughout the deformation process in the post-crushing regime; the wedge angle, $\alpha(=2\varphi)$, as measured from the specimens tested, was 100°–110° for the statically loaded shells and about 90° for the dynamically loaded ones. Normal stresses develop on the sides of the debris wedge, followed by shear stresses along the same sides, due to the friction at the interface between the wedge and the fronds. Addi-

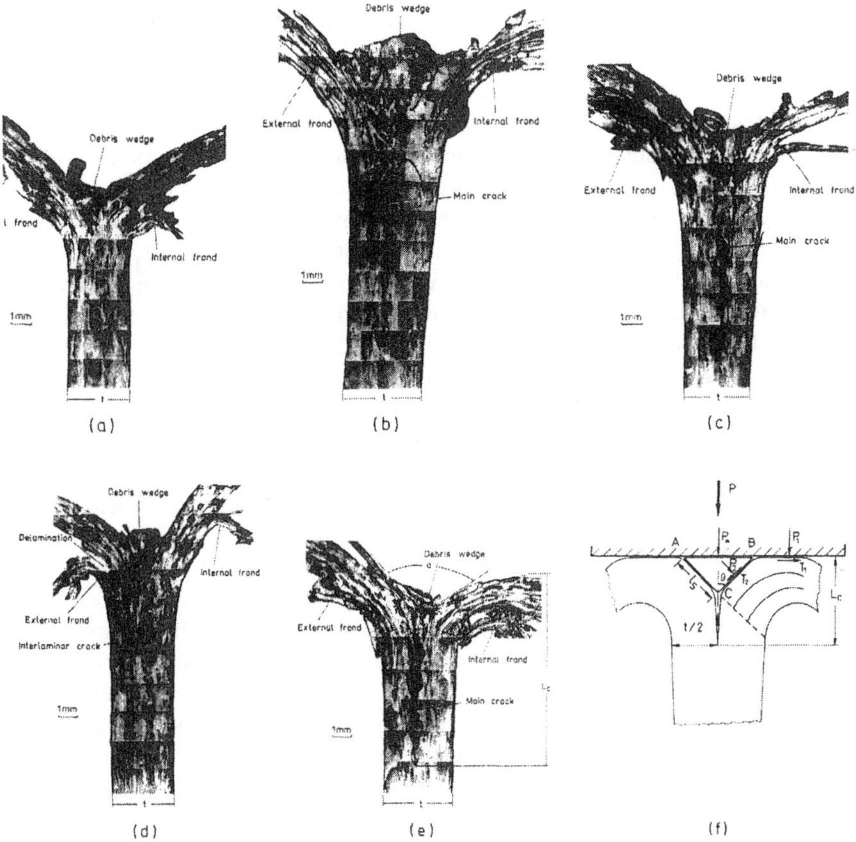

Figure 9.3. Micrographs showing microfailures in the crush-zone for dynamically loaded hourglass rail shell (sp. 18) for positions of Figure 9.1(a): (a) a, (b) b, (c) c, (d) d, (e) e, (f) configuration of the crush zone of (e).

tional normal and shear stresses develop at the interface between the steel platen or the drop mass and the deforming shell, as the formed fronds slide along this interface.

The behaviour of the reinforcing fibres depends upon their orientation. Axially aligned fibres (0°) were bent inwards or outwards, with or without fracturing, according to their flexibility and the constraints induced by other fibres, see Figure 9.2(c); their effective flexibility depends upon the fibres structure in the composite material. Fibres aligned in the hoop direction (90°) can only expand outwards by fracturing and inwards by either fracturing or buckling, see Figure 9.2(a).

Delamination occurred as a result of shear and tensile separation between plies. The axial laminae were split into progressively thinner layers, forming, therefore, translaminar cracks normal to the fibres direction, mainly due to fibre buckling; they resulted finally either in fibre fracture or in intralaminar shear cracking, splitting the laminate into a number of thin layers without fibres fracture, see Figure 9.2(c). Cracks propagate preferably through the weakest regions of the structure of the composite material, i.e. through resin-rich regions or boundaries between hoop fibres, resulting in their debonding, or through the interface between hoop and axial plies, causing delamination. Note that resistance to crack propagation along the central region of the wall is obtained by the compressive stiffness of intact internal material and by the tensile strength of the outer plies and the interply bonding.

Taking also into account the similar observations reported in Chapters 5–8, the following principal sources of energy dissipation at microscopic scale, based on the above mentioned micromechanism, as well as on secondary failure mechanisms, contributing to the overall energy absorption during collapse, may be listed: Intra-wall crack propagation; fronds bending due to delamination between plies; axial splitting between fronds; flexural damage of individual plies due to small radius of carvature at the delamination limits; frictional resistance to axial sliding between adjacent laminates; frictional resistance to the penetration of the debris wedge; frictional resistance to fronds sliding across the platen.

Relatively long specimens exhibited extensive brittle fracture designated as Mode III of collapse, see Figure 9.1(c). Fracture started at a distance from the loaded end of the specimens, approximately equal to the mid-height of the shell, and involved catastrophic failure by cracking and separation of the shell into irregular shapes, probably due to local severe shear straining of the wall of the shell. This failure mode is similar to the Euler column-buckling of very thin metallic and PVC tubes subjected to axial loading [8]. Note that the characteristic failure of the shell is usually by buckling over its weak axis of symmetry. Note, also, that for impact loading the Mode III of collapse occurred for shorter shells; compare the crushing characteristics for static and dynamic loading presented in Table 9.1.

COLLAPSE MODES AT MICROSCOPIC SCALE

The micromechanism concerning the progressive stable collapse mode (Mode I), as discussed above in detail, is the characteristic failure mechanism for the straight parts of the hourglass section, where there is no change of the specimen curvature, i.e. positions a, c and e of Figure 9.1(a), around the corners, i.e. position d, and the circular parts of the shell, i.e. position B in Figure 9.1(a), fragmentation of fibres, which are located parallel to the longitudinal axis of the shell, is the governing failure mechanism observed.

In the case of the static axial collapse, from Figure 9.2 the following microscopic observations may be listed :

- The typical microfailures observed for the straight parts of the shell are shown in Figures 9.2(a), (c) and (e). In all cases, the dimensions of the main central crack remained almost the same for both materials and independent of the shell length. Severe frond bending occurred and a debris wedge was formed, the size of which remained almost constant throughout the crushing process. Note, also, that besides the main crack, additional shear, intralaminar and interlaminar cracks developed in the straight part of the shell parallel to its longitudinal axis of symmetry; compare Figures 9.2(a) and (c) with Figure 9.2(e). Contrariwise, in the curved parts around the corners of the shell, severe fragmentation of the chip of the internal frond was observed, see Figure 9.2(b) , whilst a deeper and wider main crack was formed in the curved part B, compare Figure 9.2(b) and Figures 9.2(a), (c) and (e).

- The initial axial length of the shell, L, up to a critical length of 200 mm, seems to not affect the length of the main central crack, L_c and the mechanical response of the shell due to failure, see Table 9.1. Longer specimens exhibited different failure characteristics, i.e. shorter and more narrow main crack, whilst in the straight part, at position a of Figure 9.1(a), 2–3 small axial cracks were formed instead of the typical main central crack, with the simultaneous formation of the external frond only.
- The thickness of the specimen and the stacking conditions seem to affect the failure micromechanism. Thin specimens of 3.3 mm wall thickness showed a narrow and shallow central crack, as compared to thicker specimens of 4.3 mm wall thickness, probably due to the weaker interply bonding of the ($\pm 45°$) plies, while additional longitudinal cracks of shorter length at both sides of the main central one appeared simultaneously; compare Figures 9.4(a) and (b).

Dynamic loading greatly affects the microfracture mechanism outlined above, as it is clearly indicated from Figure 9.3. The size of the debris wedge, see also Figure 9.3(f), and the main crack dimensions, caused by the wedge propagation, are smaller

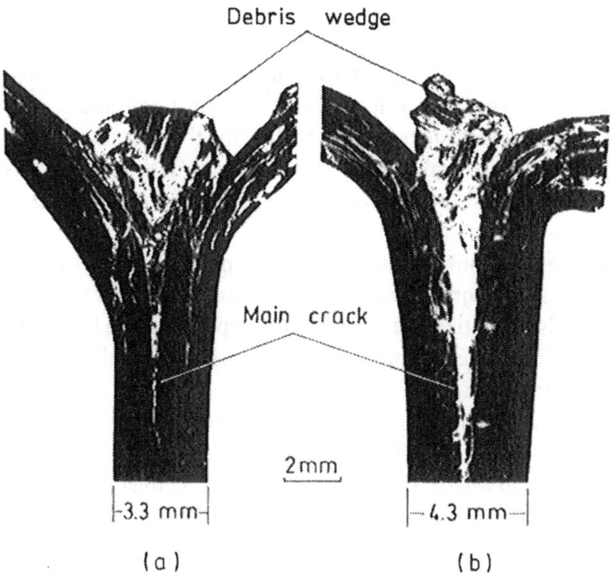

Figure 9.4. Micrographs showing the crush zone of statically loaded hourglass rail shell of wall thickness: (a) 3.3 mm, (b) 4.3 mm.

in the case of impacted shells, as compared to statically loaded ones; compare Figures 9.3 and 9.2 and see also Chapters 5 and 6. The resin behaved in a more brittle manner in dynamic loading, shattered and separated almost completely from the fibres in the crush-zone, an indication of the brittle behaviour of the composite material exhibited during its compression at elevated strain-rates. Contrary to the statically loaded shells, dynamically loaded ones exhibited side cracks, developed parallel to the main central one, compare Figures 9.3(e) and 9.2(e); these cracks are characterised by an average length considerably smaller than that of the central one. This crack formation may be attributed to the impact nature of loading requiring bigger amounts of energy absorbed in short time. Note, also, that due to the high compressive stress field developed, delamination buckling may occur without initial delamination, accompanied by an instantaneous fracture of the interlaminar bond, see Figure 9.3(d).

Comparing the micrographs of the crush-zone for the various positions a–e along the circumference of the dynamically loaded shells in Figures 9.3(a)–(e), it is evident that the main central crack did not form in the straight part of the shell, (position a) and in the corner of the shell, (position d), see Figures 9.3(a) and (d), respectively; compare also with the related statically loaded shells in Figures 9.2(a) and (d), where the development of the main central crack is apparent. Note, also, that in the position a, a characteristic irregular shape of the wedge was observed, see Figure 9.3(a), whilst in position d, fragmentation of the chip of the internal frond took place, see Figure 9.2(d); compare also with the related static tests in Figures 9.2(a) and (d).

9.3.3 Energy Absorbing Characteristics

Typical load/displacement curves for each mode of collapse for static and dynamic loading are shown in Figure 9.5. Initially the shell behaves elastically and the load rises at a steady rate to a peak value, P_{max} and then drops abruptly. At this stage, and for the Mode I of collapse, cracks form at each of the four corners accompanied by the formation of a circumferential central intrawall crack at the end of the shell adjacent to the loading area, whilst the post-crushing regime is characterised by oscillations about a mean post-crushing load, $\overline{P}$. Dynamically obtained load/displacement curves show more severe fluctuations with troughs and peaks, leading to higher energy absorbing capacity; see Figure 9.5 and also compare the static and dynamic results in Table 9.1. Note, also, that the energy absorbing capacity of the Mode III of collapse is much lower than that of the end-crushing Mode I, due to buckling of the shell and the associated brittle fracture; see also Table 9.1.

For thin-walled composite shells subjected to axial loading, it was observed that the fracture behaviour of the shell appears to affect the loading stability, as well as the magnitude of the crush load and the energy absorption, during the crushing process. It may be assumed that, at any instant, the crush load must be supported by some, more than one, structural elements of the shell and, furthermore, a specific element contributes more to supporting the load, see also Reference [81]. When this structural element fails, a subsequent structural element must then take over in supporting the load, otherwise the whole structure will fail in a catastrophic and non-energy absorb-

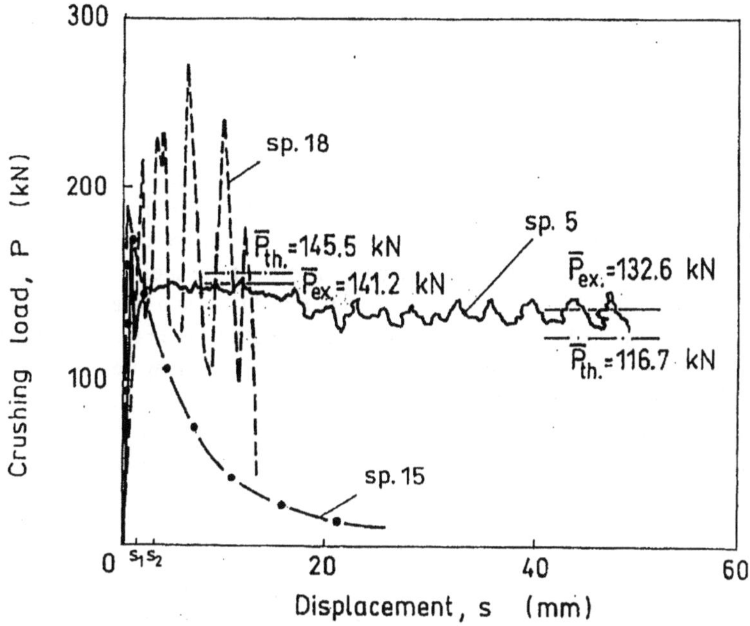

Figure 9.5. Load/displacement curves for hourglass rail shells subjected to static and dynamic loading for various collapse modes (see Table 9.1); [—— Mode I, static (sp. 5); — · — · — Mode II, static (sp. 15); – – – – Mode I, dynamic (sp. 18)].

ing manner. Therefore, during crushing, primary load support is passed from one structural element to another. This behaviour is reflected in the load/displacement curves, where the load fluctuates as a result of loading and sudden unloading, from the failure of different elements supporting the load at various times during crushing. It was shown that, for end in-plane loaded flat plates, most of the load is carried at the edges of the plate, whilst in the case of square tubes at the corners [104]. Likewise it would be expected that the present hourglass sections would support most of the load in positions around the circumference of the shell, where there is a distinct change in curvature. Therefore, the support of the surface mats depends on the integrity of the corners and the curved parts and, subsequently, they would fail before the strength of the material in the straight section would reach a critical crushing stress.

9.3.4 Failure Analysis

STATIC AXIAL COLLAPSE

The theoretical model, proposed in Chapter 5 for the analysis of composite circular tubes subjected to static axial compression, was modified and used to analyse the collapse mechanism and to estimate the related energy absorbed during the axial crushing of the hourglass sections.

Various crashworthy phenomena, pertaining to the axial collapse of composite multi-layered shells, are associated with the distribution of the absorbed energy during the crushing process. In the proposed simplified theoretical approach the following crushing phenomena were encountered: friction between the annular wedge and the fronds and between the fronds and the platen of the press; fronds bending; crack propagation; axial splitting. Note that, in the analysis, it is assumed that the crush-zone, all over the shell circumference, remains unchanged during the whole process and that axial splits occur in positions where the curvature of the cross section changes, i.e. at the four corners and the curved parts of the shell.

During the elastic deformation of the shell the load rises at a steady rate to a peak value, P_{max}, see Figure 9.5. At this stage, cracks of length L_c, see Figure 9.2(f), form at positions, where the curvature changes, and propagate downwards along the tube axis, splitting the shell wall; they are accompanied by the development of a circumferential central intrawall crack of the same length at the top end of the shell, i.e. the related shell shortening is s_1, see Figure 9.5. Therefore, the associated part of energy absorbed, which equals the external work, as can be obtained by measuring the area under the load-displacement curve in the elastic regime in Figure 9.5, is

$$W_{Lc} = R_{ad} \cdot L_c \cdot C + n \cdot (t/2) \cdot G \cdot L_c = \int_o^{s_1} P ds = \frac{1}{2} P_{max} s_1 \tag{9.1}$$

where, following the notation, R_{ad} is the fracture energy required to fracture a unit area of the adhesive at the interface between two adjacent layers, C the shell mean circumference, n the number of splits, G the fracture toughness and t the shell wall thickness.

The energy required for the deformation mechanism regarding the history of the formation of the crush-zone, see Section 5.3.4, in Chapter 5, equals the external work absorbed by the deforming shell in this regime, i.e.

$$W_{tr} = [2 \int_0^{\varphi} \sigma_o \cdot l_s \cdot (l_s/2) d\varphi] \cdot C = \int_{s_1}^{s_2} P ds \tag{9.2}$$

where, σ_o is the normal stress applied by the wedge to fronds, $l_s (= t/2 \cdot \sin\varphi)$ the side length of the wedge inscribed to the bent fronds, φ $(=\alpha/2)$ the semi-angle of the wedge, see Figure 9.2(e) and s_2 is the related shell shortening, corresponding to the completion of the wedge formation, see Figure 9.5.

Since the intrawall crack propagates at a constant speed, equal to the speed of the crosshead of the press, it can be assumed that the crack length, L_c remains constant. Also the length of the split of the crush zone (AB) at the contact side with the steel platen approximates the wall thickness, t, see Figures 9.2(e) and (f). Therefore, taking into account the failure mechanism outlined above, the total dissipated energy for a crush distance, s can be estimated as follows:

- Energy dissipated, due to friction between the annular wedge and fronds and between fronds and platen

$$W_{\mathrm{i}} = 2 \cdot (\mu_{s1} \cdot P_1 + \mu_{s2} \cdot P_2) \cdot C \cdot (s - s_2) \tag{9.3}$$

where, P_1 is the normal force per unit length applied by the platen to the internal and external fronds, P_2 is the normal force per unit length applied to the sides of the wedge, μ_{s1} is the coefficient of friction between frond and platen and μ_{s2} is the coefficient of friction between the wedge and the fronds. It must be noted that

$$P_2 = \sigma_o \cdot l_s \tag{9.4}$$

and

$$\sigma_o = k \cdot \sigma_\theta \tag{9.5}$$

where, k is a constant and σ_θ is the tensile fracture stress of the composite material.

- Energy dissipated due to fronds bending

$$W_{\mathrm{ii}} = 2 \cdot [\int_0^{\varphi} P_2 \cdot (l_s/2) d\varphi + \int_{s_2}^{s} P_2 \cdot \varphi ds] \cdot C \tag{9.6}$$

- Energy dissipated due to crack propagation

$$W_{\mathrm{iii}} = R_{\mathrm{ad}} \cdot [(s - s_1) + L_c] \cdot C \tag{9.7}$$

- Energy dissipated due to axial splitting

$$W_{\mathrm{iv}} = 8 \cdot (t/2) \cdot G \cdot s \tag{9.8}$$

From Equations (9.3), (9.4), (9.5) and (9.6) the total energy dissipated for the deformation of the shell

$$W_T = W_{\mathrm{i}} + W_{\mathrm{ii}} + W_{\mathrm{iii}} + W_{\mathrm{iv}} \tag{9.9}$$

is

$$\begin{aligned} W_T = {} & [1/(1 - \mu_{s1} + \mu_{s1} \cdot s_2/s)] \cdot [C \cdot t \cdot k \cdot \sigma_\theta \cdot \{(s - s_2) \\ & \cdot [\mu_{s2}/\cos(\alpha/2) - \mu_{s1} \cdot [\tan(\alpha/2) + \mu_{s2}]] \\ & + [(\alpha/2)/\cos(\alpha/2)] \cdot [0.25 \cdot t/\cos(\alpha/2) + s - s_2]\} \\ & + R_{\mathrm{ad}} \cdot C \cdot (s - s_1 + L_c) + 8 \cdot (t/2) \cdot G \cdot s] \end{aligned} \tag{9.10}$$

whilst the total normal force applied by the platen to the shell can be calculated as

$$P = W_T / s \tag{9.11}$$

From the stress/strain curve (material B), shown in Figure 5.4 of Chapter 5, the tensile fracture stress, σ_θ of the materials was estimated. The static friction coefficients, μ_{s1} and μ_{s2} were obtained by employing the curling test [41]. The fracture toughness, G was estimated from the tension test of notched strips. The interfacial fracture energy, R_{ad} was calculated from Equation (5.2) of Chapter 5, using the experimental results obtained by loading cylindrical tubes up to the maximum load, P_{max} and obtaining the energy absorbed from the related experimental load/displacement curves of loaded shells; R_{ad} may be also estimated by employing the strip peel test, [46].

Experimentally obtained values for μ_{s1}, μ_{s2}, R_{ad}, G, σ_θ and the constant k, for the composite material B, are presented in Table 5.2 of Chapter 5.

THE EFFECT OF STRAIN-RATE

The effect of the strain-rate on the microfailure mechanism of axially loaded tubes of various geometries is discussed in detail in Chapters 5 and 6. The microscopic differences in the crush-zones of statically and dynamically loaded thin-walled shells, see Figures 9.2 and 9.3, may well explain, qualitatively, the strain-rate effect but it renders difficulties to provide an accurate estimation based only on them. Therefore, the effort may be focused also on the effect of the various crashworthy phenomena, which affect the total energy dissipated during the crushing process, as discussed above.

The major part of the energy absorbed during the static axial compression of a shell is dissipated as frictional work in the crushed material, or at the interface between material and tool, and this is estimated to be about 50% or more of the total work done; see also Chapters 5 and 6. In dynamic collapse, attention is directed towards the influence of strain-rate on the frictional work absorbed during the impact, taking into account all structural and material parameters, which may contribute to it, i.e. fibre and matrix material, fibre diameter and orientation in the laminate, the fibre volume content as well as the conditions at the interfaces.

As mentioned above, two distinct regions, where the development of frictional forces is of great importance, were identified: the fronds/wedge contact region, composed of the same material, and the fronds/platen contact area, composed of different materials. The coefficient of friction in these regions depends on various phenomena associated with the material flow [105], such as:

- Elastic and/or plastic deformation occurring at the contact area of the sliding surfaces subjected to external load; in general, plastic deformation results in a decrease of the coefficient of friction due to the reduction of the shear resistance at the sliding surfaces.
- Interfacial bonding due to the electrostatic forces developed in the contact area; it is greatly affected by the conductivity of the materials and the temperature field developed, leading to an increase of the friction coefficient.

- Adhesion occurring at the contact region during the sliding of the two deformable bodies of the same material and, resulting in an increase of the coefficient of friction.

For the static axial collapse, sliding frictional conditions prevail with a friction coefficient μ_s. At impact a dynamic coefficient μ_d is defined as

$$\mu_d = \gamma_x / g(1 + \gamma_y / g) \tag{9.12}$$

where, γ_x and γ_y are the horizontal and normal components of the acceleration in the contact area, respectively.

Adhesion is the governing phenomenon in the frontal wedge contact area, see Figure 9.2(f), whilst interfacial bonding dominates at the fronds/platen interface.

Impact loading results in high electrostatic forces in the fronds/wedge and fronds/platen contact regions, resulting, therefore, in greater values of the dynamic coefficient of friction, μ_{d2} and μ_{d1} as compared to the static ones, μ_{s2} and μ_{s1}, respectively. In addition, the temperature in the crush-zone usually increases in the case of impact loading, leading, therefore, to higher values of the dynamic coefficients of friction as compared to the static ones.

Note, however, that the elastic/plastic deformation imposed in the sliding interfaces and the associated surface changes are not clearly defined and it is, therefore, difficult to estimate the prevailing nature, static or dynamic, of the phenomenon occurring. Knowledge, therefore, of the conditions prevailing, which are not uniquely defined during the static or dynamic axial collapse, leads to an estimate of the mechanical response and the crashworthy behaviour of the structural component during the crushing process. The governing phenomena, outlined above, are greatly affected by the material properties, influencing finally the effectiveness of the collapsed component, as far as its crashworthy capacity is concerned.

The total energy absorbed during the dynamic axial collapse of hourglass cross-sectioned shells may be, therefore, estimated from Equation (9.10) by substituting only the static friction coefficients, μ_{s1}, and μ_{s2} with the dynamic ones, μ_{s1}, and μ_{s2}, respectively, as

$$\begin{aligned} W_T = {} & [1/(1 - \mu_{d1} + \mu_{d1} \cdot s_2/s)] \cdot [C \cdot t \cdot k \cdot \sigma_\theta \cdot \{(s - s_2) \\ & \cdot [\mu_{d2}/\cos(\alpha/2) - \mu_{d1} \cdot [\tan(\alpha/2) + \mu_{d2}]] \\ & + [(\alpha/2)/\cos(\alpha/2)] \cdot [0.25 \cdot t/\cos(\alpha/2) + s - s_2]\} \\ & + R_{\text{ad}} \cdot C \cdot (s - s_1 + L_c) + 8 \cdot (t/2) \cdot G \cdot s] \end{aligned} \tag{9.13}$$

An estimate of the dynamic coefficients of friction at the fronds/platen interface, μ_{s1} and the fronds/wedge intertace, μ_{s2} was obtained by comparing the experimental crushing loads in Table 9.1. It must be noted that these values equal the corresponding ones of circular tubes of the same material (material B), presented in Table 5.2 of Chapter 5.

9.3.5 Crashworthy Capability: Concluding Remarks

Two collapse modes were observed; the stable progressive collapse mode, Mode I, associated with large amounts of crush energy, resulting, therefore, in a high crashworthy capacity of the structural component in head-on collisions and the mid-length collapse mode, Mode III, with rather low energy absorption capability.

The microfracture mechanism observed for the progressive collapse Mode I, is, in general, similar for statically and dynamically loaded shells, respectively. The only differences encountered are associated with the shape of the wedge and the microcracking development.

The depth, to which the cracks penetrate due to axial splitting at the corners, see Figures 9.2(b) and 9.3(d), seems to be related to the material lay-up used, to the possible residual stress field developed during the manufacturing process and to the type of triggering used. However, for the specimens tested triggering was not employed. As far as the residual stresses at the corners is concerned, it was observed, that delamination in this region is not entirely attributed to residual circumferential or axial stresses existing there [113]. Therefore, the differences in crashworthy characteristics encountered for the statically loaded shells, made from the two different types of lay-up (materials A and B), are propably due to the existence of the (±45°) central plies in the material B. Thus, because of the deeper crack penetration that resulted by the material B lay-up, see Figure 9.4, the load supporting ability of the shells and the specific energy absorbed are reduced, see also Table 9.1. This agrees with similar remarks reported in References [114, 115] where cohere tubes with a $(45/45)_n$ lay-up developed consistently lower values of specific energy than tubes with $(0/90)_n$ lay-ups in the stable collapse region.

From Table 9.1, it is evident that the mean post-crushing load and the energy absorbed are mainly affected by the crush length, whilst the axial length of the shell has no significant effect on these crashworthy characteristics. Note, also, that higher values of mean post-crushing load and energy absorbed were predicted for dynamic collapse, see Figure 9.5 and Table 9.1, probably due to the higher values of the dynamic friction coefficients at the interfaces between the wedge and the fronds and between the drop mass and the fronds (about 15–20% of the related static ones); this increase in the crashworthy ability of the shell was about 20%.

The mean post-crushing load, $\overline{P}$ and the energy absorbed, W are well predicted theoretically by the proposed analysis, within ±10%, see Table 9.1. Note that in this approach, fibre orientation was not taken into account. According to the proposed theoretical analysis the distribution of the dissipated energy of the crush shell, due to the four main energy sources, was estimated as:

- Energy due to friction between annular wedge and fronds and between fronds and platen, about 45% of the total one
- Energy due to fronds bending, about 40%
- Energy due to crack propagation, about 12%
- Energy due to axial splitting at the four corners and the four circular parts of the shell, about 3%

The contribution of the frictional conditions between wedge/fronds and fronds/platen to the energy absorbing capability is more significant than the other ones. As discussed above, it mainly depends upon the friction coefficients μ_{s1}, μ_{d1} and μ_{s2}, μ_{d2}, which are affected by the surface conditions at the interfaces between com-

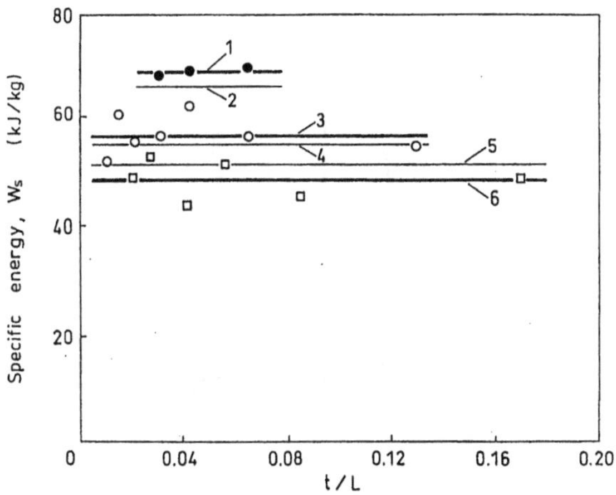

Figure 9.6. Variation of specific energy, W_s, with ratio *t/L* for: material A dynamically loaded 1: ● experimental, 2: —— theoretical; material A statically loaded 3: ○ experimental, 4: —— theoretical, material B statically loaded 5: □ experimental, 6: — theoretical.

posite material/platen or drop mass and composite material/debris wedge, respectively. From the analysis, it is also evident that the annular debris wedge supports the 55% of the crush load and the internal and external fronds the remaining 45%, see Table 9.1; this is in agreement with similar remarks reported in Chapters 5 and 6. It was also observed, that the energy absorbed due to bending of the fronds is greatly affected by the magnitude of the wedge semi-angle, which varies between 45° and $\varphi = \tan^{-1}(1/\mu_2)$.

In Figure 9.6 experimental values of the specific energy, W_s are plotted against the geometry factor thickness/axial length, *t/L*, of the shell, along with the related theoretically predicted ones, see also Table 9.1. The specific energy seems to be almost constant, supporting, therefore, the remarks made above that, for a constant thickness of the shell, its axial length has no significant effect on its energy absorbing capability. Dynamic values overestimate the static ones for the same meterial by about 20%, whilst theoretical and experimental values are in good agreement to within ±10%.

9.4 BENDING

9.4.1 Experimental

Bending crush tests were carried out to observe the collapse rail beams of non-conventional cross-sectional geometry, as shown in Figure 9.1(a). This structural component constitutes the lower rail structure of the composite car front end.

Three rail beams were cantilevered and bent about their strong axis, under various constraints at their ends, through a plug inserted individually at each of the beam ends over a length of 100 mm, see Table 9.2 for details; two more rail beams were bent

Table 9.2: Crushing characteristics of hourglass cross-sectioned shells subjected to bending

Specimen no	Bending conditions	Thickness t (mm)	θ_{max} (°)	Peak moment, M_{max} (N m)		Energy absorbed W_T(J)	Testing conditions
				Experimental	Theoretical		
1	Bent over strong axis	2.80	11.5	2820.0	2791.8	316.4	No plug
2	"	2.80	11.5	2784.0	2791.8	480.9	Plug at supported end
3	"	2.80	8.0	2862.0	2791.8	495.1	Plug at clamped end
4	Bent over weak axis	2.80	10.0	1132.0	1156.2	178.9	No plug
5	"	2.80	11.0	1150.0	1156.2	238.5	No plug

about their weak axis without end-constraints. The span length was kept constant and equal to 250 mm.

The experimental set-up is described in detail in Section 5.4.1 of Chapter 5, and only a short description is given here. The torque required for the tube bending was supplied by a speed reducer driven by an electric motor. The specimen was suitably clamped at one end and supported at a point close to its opposite end. The tube holding fixture was attached to the output shaft of the speed reducer, providing in this manner with the tube rotation. The whole process represents the bending of cantilever beams. The loading system was equipped with measuring devices, which through a data-acquisition system provided with the bending moment (I)/angle of rotation (θ) curves during the bending process.

The material used was a commercial fibreglass and vinylester resin composite material, designated as material B. The tube wall consisted of 9 plies which have a total thickness of 3.3 mm. From the outside towards the inside surface of the tube, the plies are laid-up in the sequence $[(90/0/2R_c)/(2R_c/0/90)/R_{c.75}]$, where the 0°direction is along the axis of the tube; R_c denotes the random chopped strand mat plies and $R_{c.75}$ represents a similar ply but thinner. According to the manufacturer's specifications, the laminate fibre lay-up was hand wrapped around a rigid foam core and plastic staples were used to hold the lay-up in the core during the resin transfer molding process. The specimens were injection molded and allowed to cure for 45 min at room temperature and, then they were post-cured at 120°C for 3 *h*. Finally, the tubes were defoamed and cut properly into 400 mm lengths. More details in the material properties and the laying-up are given in Section 5.4.1 of Chapter 5.

Tensile and compression tests were performed in an Instron testing machine at a low crosshead speed of about 10 mm/min. All test results concerning the strength characteristics of the material are tabulated in Table 9.3, whilst a typical stress-strain curve (material B) under tension is given in Figure 5.4 of Chapter 5.

Table 9.3: Material properties of the plies in the laminate

	Unidirectional plies	Random chopped mat plies	Laminate
E_1 (GPa)	7.7	10.6	12.7
E_2 (GPa)	10.3	10.6	18.4
G_{12} (GPa)	4.1	4.4	
v_{12}	0.33	0.336	0.151
v_{21}	0.09	0.336	0.140
X_t (GPa)	0.872	0.116	0.192
X_c (GPa)	0.605	0.116	0.236
Y_t (GPa)	0.034	0.116	0.332
Y_c (GPa)	0.129	0.116	0.335
S (GPa)	0.139	0.117	

A series of photographs of crushing modes were taken during the various loading stages, which, along with terminal macroscopic views of the specimens tested and the associated M/θ curves, are shown in Figures 9.7–9.11. The experimental results, pertaining to loading and energy absorbing (i.e. the area under the M/θ curve) characteristics, are tabulated in Table 9.2.

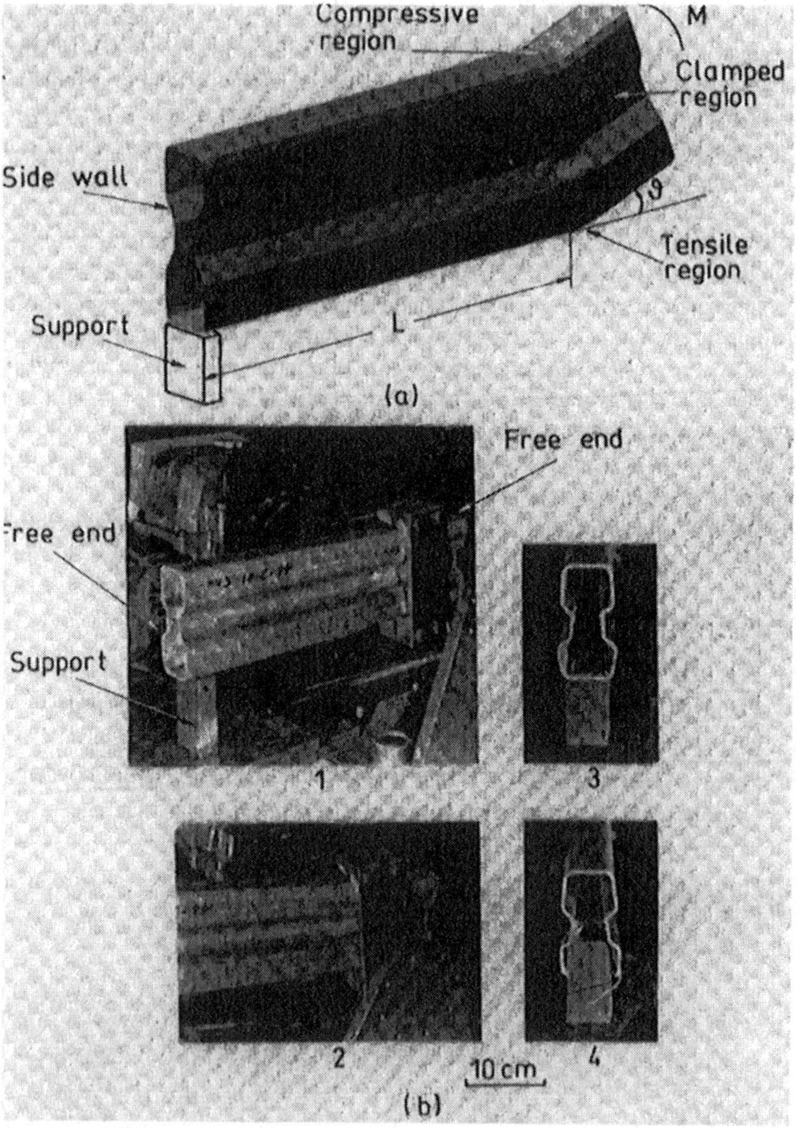

Figure 9.7. (a) Bending of a rail beam over its strong axis, (b) progressive collapse of specimen 1 (see Table 9.2).

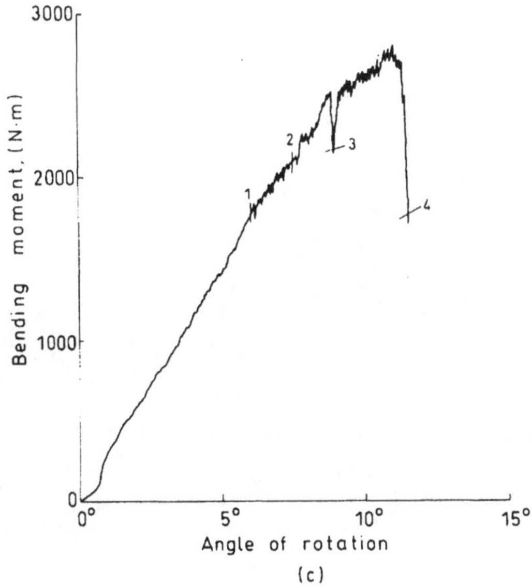

Figure 9.7 (continued). (c) Bending moment/angle of rotation curve for specimen 1 (numbers 1–4 refer to the deformation stages of Figure 9.7(b)).

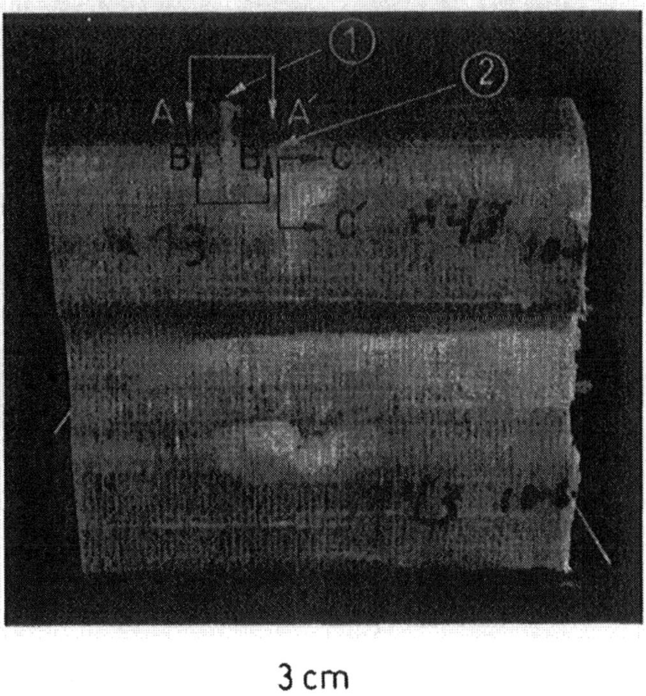

Figure 9.8. Macroscopic side view of specimen 1 (see Table 9.2) after collapse.

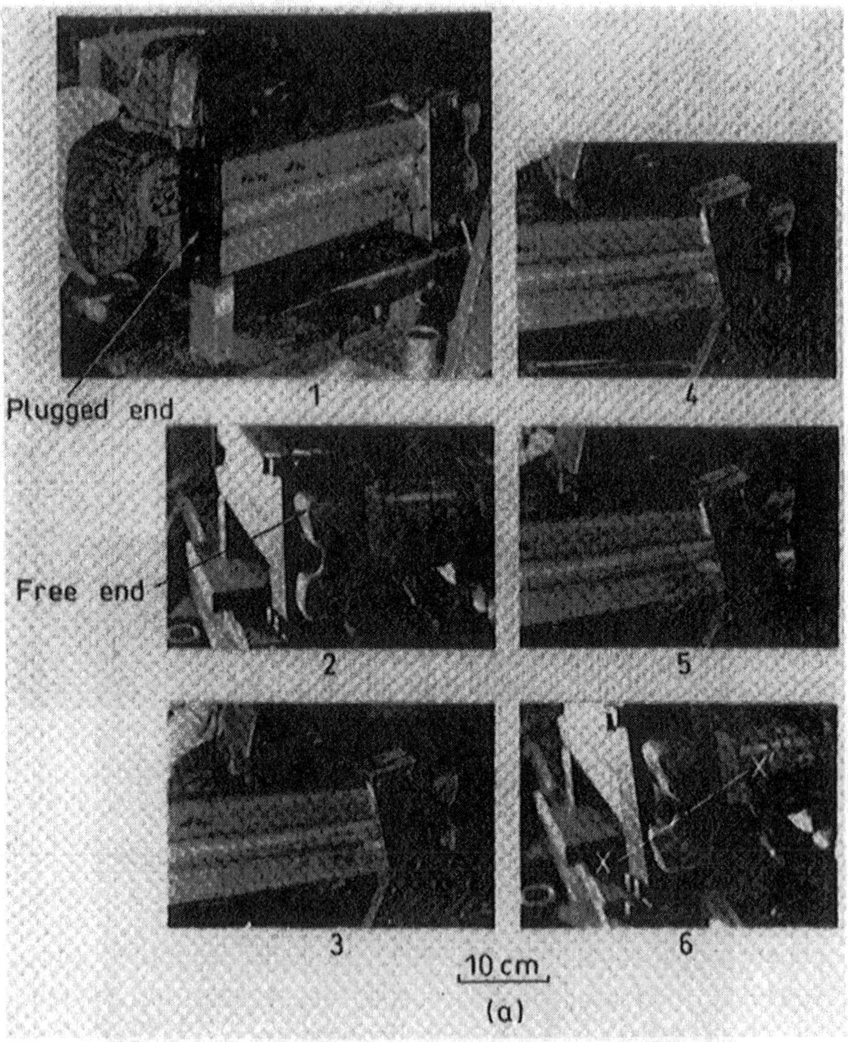

Figure 9.9. (a) Progressive collapse of specimen 2 (see Table 9.2).

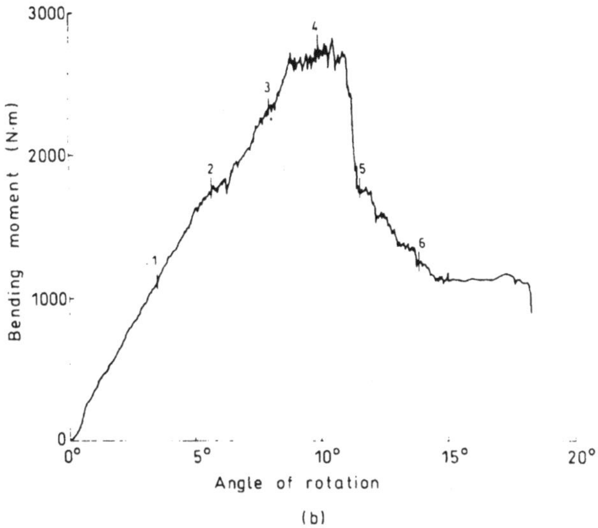

Figure 9.9 (continued). (b) Bending moment/angle of rotation curve of specimen 2 (numbers 1–6 refer to the deformation stages of Figure 9.9(a)).

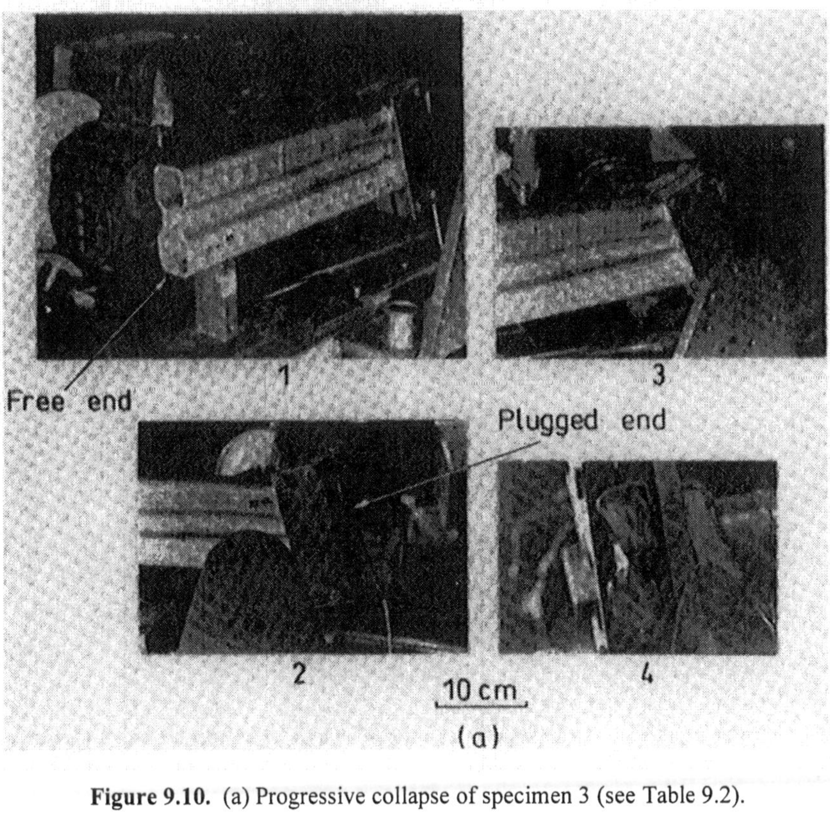

Figure 9.10. (a) Progressive collapse of specimen 3 (see Table 9.2).

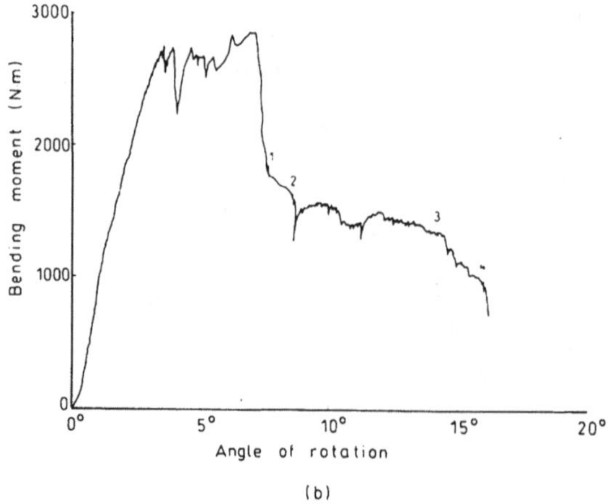

Figure 9.10 (continued). (b) Bending moment/angle of rotation curve of specimen 3 (numbers 1–4 refer to the deformation stages of Figure 9.10(a))

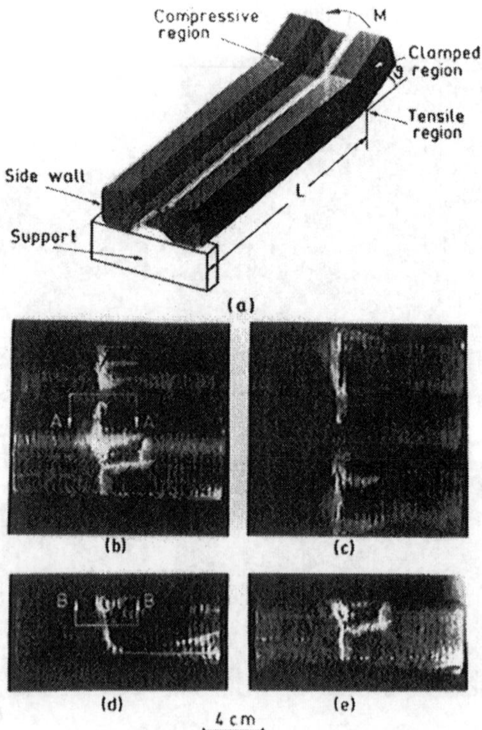

Figure 9.11. (a) Bending of a rail beam over its weak axis. Macroscopic views of specimen 4 (see Table 9.2) after collapse: (b) compressive zone, (c) tensile zone, (d) front side, (e) back side.

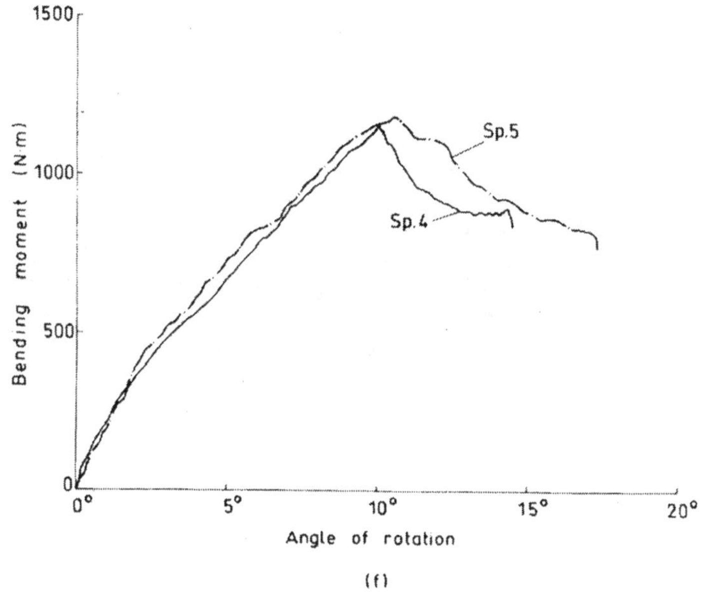

Figure 9.11 (continued). (f) Bending moment/angle of rotation curves for specimens 4 and 5.

Sections, cut from crushed regions of the tube wall, were properly prepared and polished in six stages, ranging from 200 grit abrasive wheel to 0.25 μm alumina paste, see also Section 5.3.1 of Chapter 5 for details. These metallographic specimens were then examined using a Unimet metallographic optical microscope. Typical micrographs showing the fracture patterns at various positions of the crushed zone are presented in Figures 9.12–9.19.

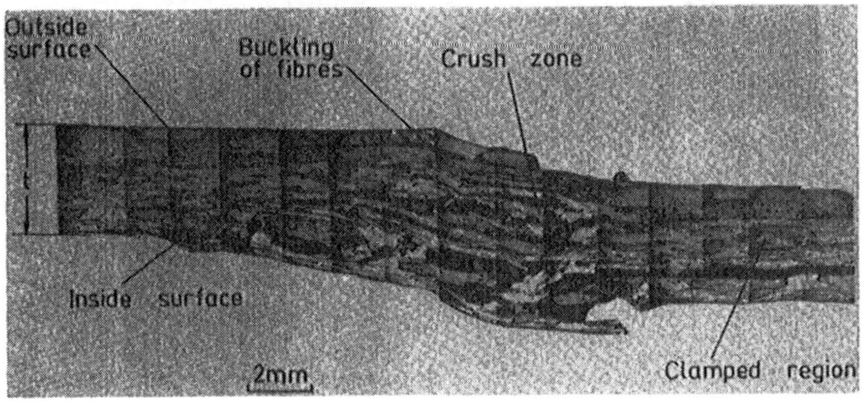

Figure 9.12. Micrograph at section AA′ of Figure 9.8 showing microfailures of the compressive zone of specimen 1.

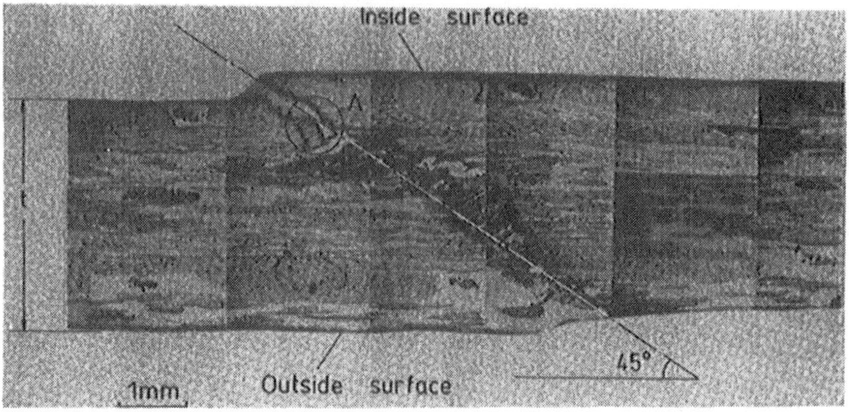

Figure 9.13. Micrograph at section BB′ of Figure 9.8 showing microfailures of the side wall of specimen 1.

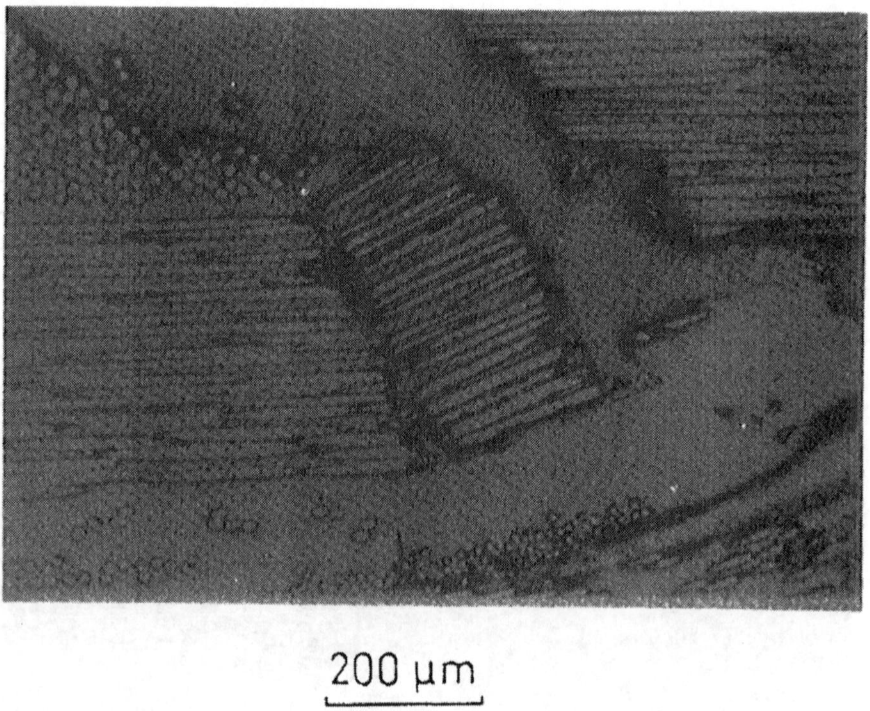

Figure 9.14. Magnification of region A of Figure 9.13 showing typical microfractures in the side wall.

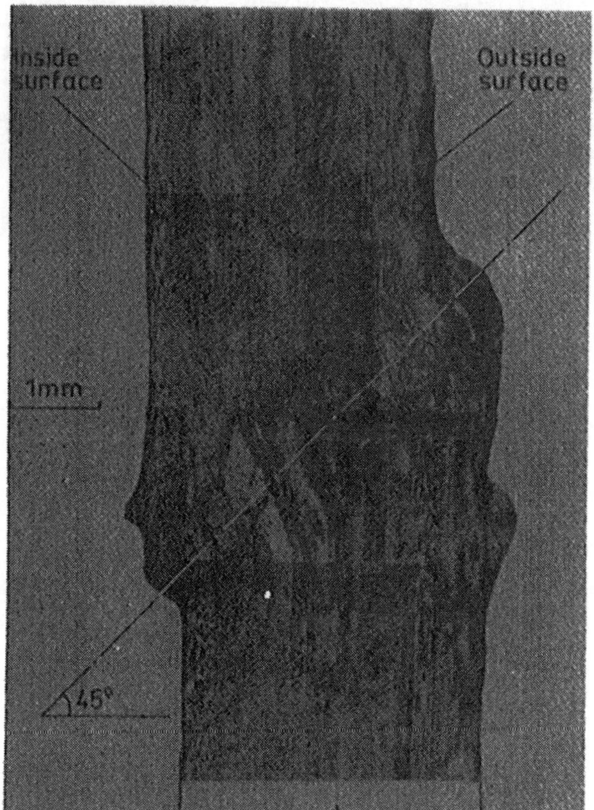

Figure 9.15. Micrograph at section CC′ of Figure 9.8 showing microfailures of the side wall of specimen 1.

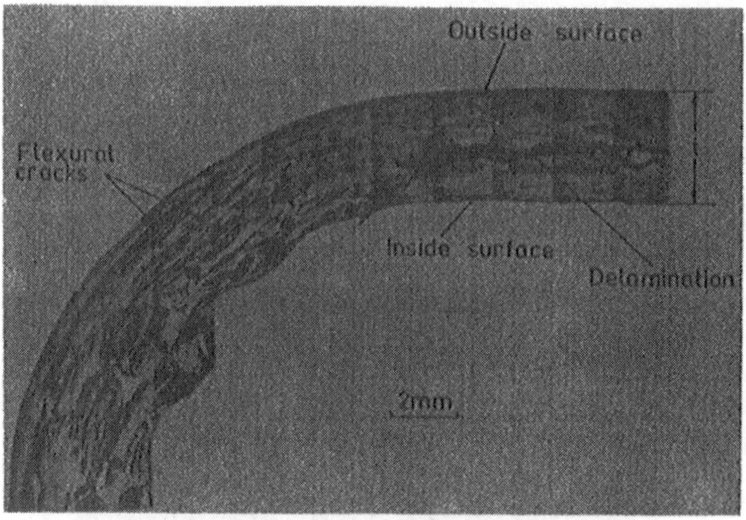

Figure 9.16. Micrograph showing microfailures at the corner of position 1 of Figure 9.8.

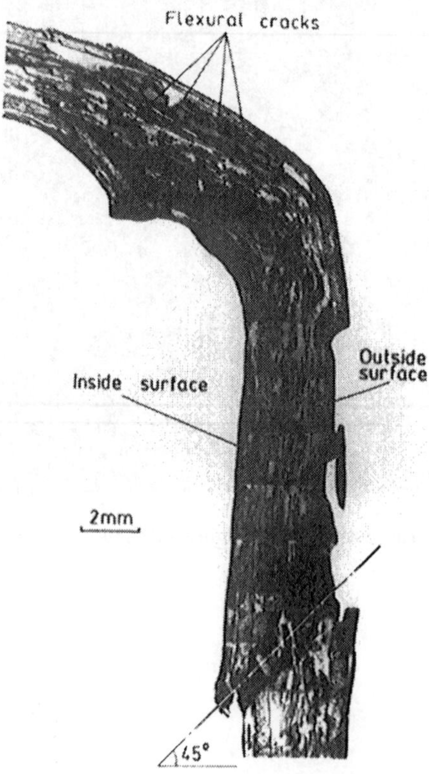

Figure 9.17. Micrograph showing microfailures at the corner of position 2 of Figure 9.8.

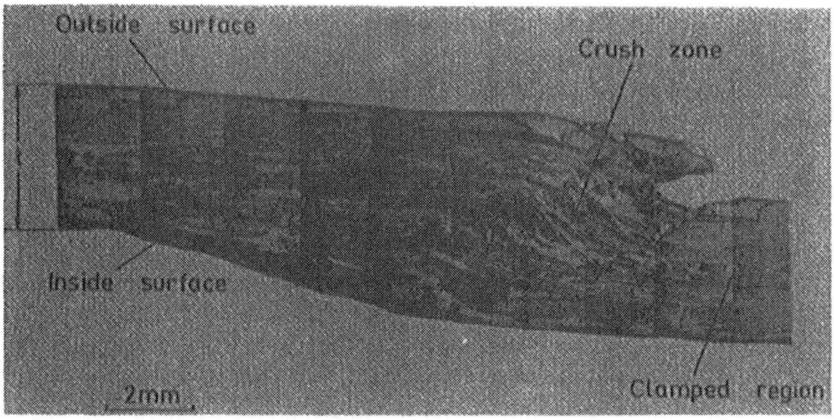

Figure 9.18. Micrograph at section AA′ of Figure 9.11(b) showing microfailures of the compressive zone of specimen 4.

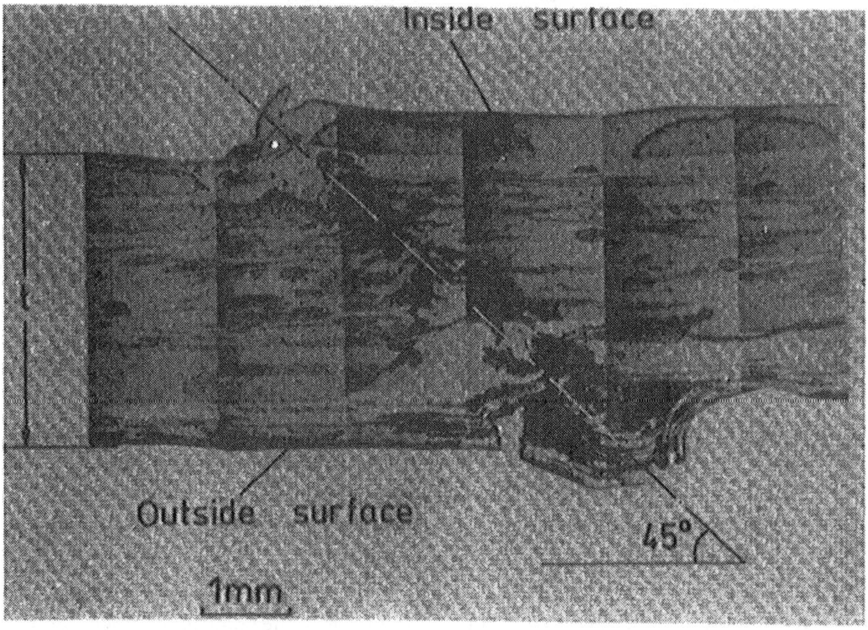

Figure 9.19. Micrograph at section BB′ of Figure 9.11(d) showing microfailures of the side wall of specimen 4.

9.4.2 Failure Mechanisms

MACROSCOPIC DEFORMATION MODES

In almost all cases of the bent thin-walled sections examined, several common deformation and strength characteristics are apparent, see Figures 9.7–9.11:

- Three distinct regions were observed with different straining features: the top compressive side, the bottom one subjected to tensile straining and the intermediate part under combined compression/tension.
- The predominant macroscopic failure mode is extensive fracturing in the compression zone, adjacent to the edge of the clamping device. For bending about the major axis, failure occurred at the top compression zone, but the bottom tension zone experienced no failure, even after the contact region at the loading zone collapsed entirely. Cracking along the middle surface was also observed, probably due to the maximum shear stress developed in this region. For bending about the minor axis, failure also occurred at the top compression zone and the top corners were sheared, while the tensile zone remained intact. No cracking was observed along the middle surface.

The effect of the plug insert on the deformation history of the tubular component may be described as follows:

- Specimen 1, with no plugs at its free edges, see 1 in Figure 9.7(b), initially behaved elastically up to an angle of rotation of about 7°, then a severe serration of the load-deflection curve developed, probably due to the start of microcracking of the most highly strained plies of the compression zone, see Figure 9.7(c). The fracture region spread quickly from the axis of symmetry of the top side towards the corners. A large number of cracks developed, just underneath the clamping edge, and, as the hinge rotation continued, the tube sustained higher loads as the corners began to crush, whilst the compressive top side and the side walls started to collapse. Simultaneously, flattening of the lower portion of the cross section at the beam support was observed, see 3 in Figure 9.7(b), leading abruptly to local bending and rupture of the bottom plate of the tube, see 4 in Figure 9.7(b). After this phase, no further loading capacity is available, see Figure 9.7(c). The terminal fracture after collapse is shown in Figure 9.8.
- Specimen 2, possessing a strengthening plug at its free supported end with the clamped end in excess over a length of 20 mm from the clamping device, see 1 in Figure 9.9(a), did not develop the above-mentioned local rupture; see 2–5 in Figure 9.9(a). Up to the maximum strength, the same deformation characteristics, as mentioned previously, were observed but a further load-carrying capacity was attained in the post-crushing region over a wide range of the hinge rotation, see Figure 9.9(b). It must be noted that, at the end of the deformation, a severe distortion and fracture of the clamped cross section

was observed, leading to disturbed clamping conditions and a loss of the tube load carrying capacity; see 6 in Figure 9.9(a).

- Specimen 3, with a plug at its clamped end and the supported end extending about 20 mm from the support, initially sustained higher loads than the specimens 1 and 2, developing an increase of the slope of the elastic portion of the load/deflection curve, up to the maximum strength, and keeping this value for a relatively wide range of hinge rotation, see Figures 9.10 (a) and (b). In the post-crushing regime the specimen reserved its load-carrying capacity until the strengthening plug was expelled, probably due to the severe cross-sectional distortion of the clamped end, see 4 in Figure 9.10 (a). No local deformation and rupture of the supported end was observed.
- Specimens 4 and 5, bent about their weak axis under the same loading and clamping conditions, free of strengthening plugs, showed quite repeatable deformation patterns and load-carrying capacity characteristics, see Figures 9.11 (a)–(f).

MICROSCOPIC OBSERVATIONS

From the micrographs obtained in the crushed regions of the bent specimens, the following microscopic observations are made, see Figures 9.12–9.19:

- Compressive failure at the top side is characterised by slipping and/or fracturing of material. The delamination mechanism is usually activated. Small internal cracks, perpendicular to the longitudinal axis of symmetry of the tube, are found close to the region of maximum deformation and are accompanied by similar cracks of random orientation, corresponding mainly to whitened areas of the deformed zone, see Figures 9.12 and 9.18 as well as the macroscopic views of the bent specimens in Figs 9.7 and 9.11. A central crack usually forms, which propagates longitudinally through the wall thickness. This crack is bifurcated to smaller ones by crossing regions with small fibre volume fraction or by growing through the boundaries between fibres of different orientation.
- From the failure surface topography of the side walls, see Figures 9.13–9.15 and 9.19, a complex damage mechanism can be suggested consisting of the combination of the following straining conditions: longitudinal tensile failure, causing cracking normal to the fibres, leading to fracture or debonding of fibres and pull-out of broken fibres from their sockets in the resin; longitudinal compressive failure, caused by the combined action of the following superimposed secondary collapse mechanisms: (a) elastic local microbuckling of the composite material, mainly in the areas with low volume fractions and (b) shear failure of resin and fibres through a kink band, progressing across the specimen; transverse tensile failure, inducing cracking in the longitudinal direction; transverse compression failure, leading to collapse under an inter- or intralaminar shear mechanism.
- The regions close to the corners are subjected to complex straining, resulting in uncontrolled fracture patterns, see Figures 9.8, 9.16, and 9.17, accompanied by

severe cracking of the neighbouring areas, seriously affecting the buckling behaviour of the compression zone of the tubular component.

In general, the main cracking modes, developed in the various damage areas, may be classified as, flexure, delamination buckling and shear mode, depending on loading conditions (about the strong or the weak axis), the strained region (compression zone, corner or side wall) and cracking development.

- In flexural mode the typical damage was fibre and matrix breakage on either side of the laminate (compression zone). The damage is limited to a small region at the centre of the thickness, with secondary cracks extending further on the plate surface or propagating through the fibre bundles. On the concave side, the 90° fibres failed primarily under compressive straining, resulting in a wide debonded, whitened area and a loss of the lateral support of the 0° plies, which primarily bent without damage to the fibres. Some of the broken fibres and matrix protruded from the surface. On the convex side, the random chopped fibre mat or the 90° fibres show a tendency for pulling-out.
- Delamination buckling mode was the dominate failure mode. A delamination crack was developed, usually in the middle of the laminate, extended at both directions (x-y) of the laminate and towards the clamping device as the reflection increases. Crack propagation occurred more or less in the interphase between the two random chopped fibre mats in the middle of the laminate (corner, side walls). The main features of this mode of collapse were: fibre and matrix breakage at the centre of the laminate; delamination cracking, propagating through the interphase between the random chopped fibre mat or the interphase between 0° and 90° fibres without much fibre breakage; bending of the 0° fibres without damage to the fibres.
- In the failure by shear mode, the typical damage was fibre and matrix breakage through the thickness of the specimen. For crushed zones developing no delamination, the damage was primarily limited to a small region, forming a shear band at an angle of 45° to the beam axis. There were no secondary cracks on the surfaces. The random chopped fibre mats failed and twisted, whilst the 0° fibres were broken due to the kink band formation. For regions with delamination, in addition to the shear band occurred on the convex side, a delamination crack through a wide area was formed on the concave side of the laminate. The 0° plies were only bent with no damage to the fibres. In both cases, the 90° fibres usually broke.

9.4.3 Energy Absorbing Characteristics

Initially the shells behave elastically. A steady state increase of small slope of the bending moment, see Figures 9.7(c), 9.9(b), 9.10(b), and 9.11(f), corresponds to the elastic bending, accompanied by extensive microfracturing in the shell wall in the compressive region. Then, after a small hinge rotation, θ, ranging between 2°– 7°, the slope of the M/θ curve changes, accompanied by a small serration with increasing moment up to a peak value, M_{max}. During this phase, the load is carried by the compressive top side and the side walls of the tube, due to buckling and crushing in the extended region of the hinge rotation. As the maximum moment is attained, after a considerable amount of hinge rotation, the edges undergo severe material fragmentation, leading to a decrease of the bending moment. Deep collapse follows, where the fractured edges interact with each other, causing further resistance to bending of the shell, strongly depended upon the plug insert.

Rail beams, with a strengthening plug at their supported end, show the same load/deflection characteristics in the elastic regime of deformation with the unplugged shell, but a further energy absorbing capacity was attained in the post-crushing region over a wide range of the hinge rotation; compare Figures 9.7(c) and 9.9(b).

In the case of rail beams with a strengthening plug at their clamped end, an increase in the slope of the elastic portion of the load/deflection curve, up to the maximum strength, was attained; this value was kept constant over a wide range of hinge rotation, compare the curves in Figures 9.7(c) and 9.10(b). An adequate energy absorbing capacity in the post-crushing regime is also apparent for this end-constraint.

Beams, bent over their weak axis, developed much lower strength and energy absorbing ability, compared to those bent over their strong axis, see Figures 9.7(c) and 9.11(f).

9.4.4 Failure Analysis

The classical lamination theory was suitably modified to cover the deformation mechanism of a square or rectangular tube, see Section 6.4.4 of Chapter 6. This failure analysis may be also valid in the case of the present shell geometry with the appropriate modifications.

Consider the cantilever laminated composite tube of Figure 9.20, subjected to an end-shear load, P. If X-Y is the coordinate system applied to the beam, then, according to the simple theory of flexure, the state of stress in the compression zone for the lamina i may be expressed as:

$$\begin{bmatrix} \sigma_y \\ \sigma_y \\ \tau_{xy} \end{bmatrix} = \begin{bmatrix} \pm PLY^i / l_{XX} \\ 0 \\ 0 \end{bmatrix} \tag{9.14}$$

where, I_{XX} is the moment of inertia of the tube cross section.

The corresponding solution for the indermediate side walls, being under combined compression/tension straining, gives

$$\begin{bmatrix} \sigma_x \\ \sigma_y \\ \tau_{xy} \end{bmatrix} = \begin{bmatrix} \pm PLY^i / l_{XX} \\ 0 \\ PD^2[1-(2Y^i/D)^2]/8I_{XX} \end{bmatrix} \tag{9.15}$$

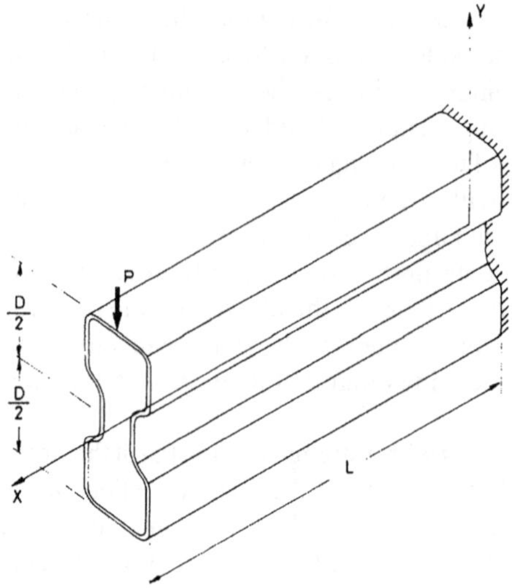

Figure 9.20. Bending of a cantilever rail beam.

where, D is the height of the beam cross-section, see also Figure 9.20. The prediction of the ultimate strength requires the selection of an appropriate failure criterion. In the present case, the 2-dimensional Tsai-Wu failure criterion [52] is used, expressed in the form

$$F_{11}\sigma_1^2 + F_{22}\sigma_2^2 + 2F_{12}\sigma_1\sigma_2 + F_{66}\sigma_6^2 + F_1\sigma_1 + F_2\sigma_2 + F_6\sigma_6 = 1 \qquad (9.16)$$

where,

$$F_{11} = 1/X_t X_c$$
$$F_{22} = 1/Y_t Y_c$$
$$F_{66} = 1/S^2$$
$$F_1 = 1/X_t - 1/X_c$$
$$F_2 = 1/Y_t - 1/Y_c$$
$$\sigma_6 = \tau_{xy}$$
$$F_6 = F_{12} = 0$$

X_t, Y_t are the uniaxial tensile strengths of the lamina i in the longitudinal and transverse direction, respectively and X_c, Y_c the related uniaxial compressive strengths. S is the shear strength of this lamina.

The stresses, predicted from Equations (9.14) and (9.15), are then compared with the failure strength of the material to determine whether a particular layer has failed. After a lamina fails, the procedure is repeated until total laminate rupture. At any stage of loading, the vertical deflection of the tube at the loaded end is given by the equation:

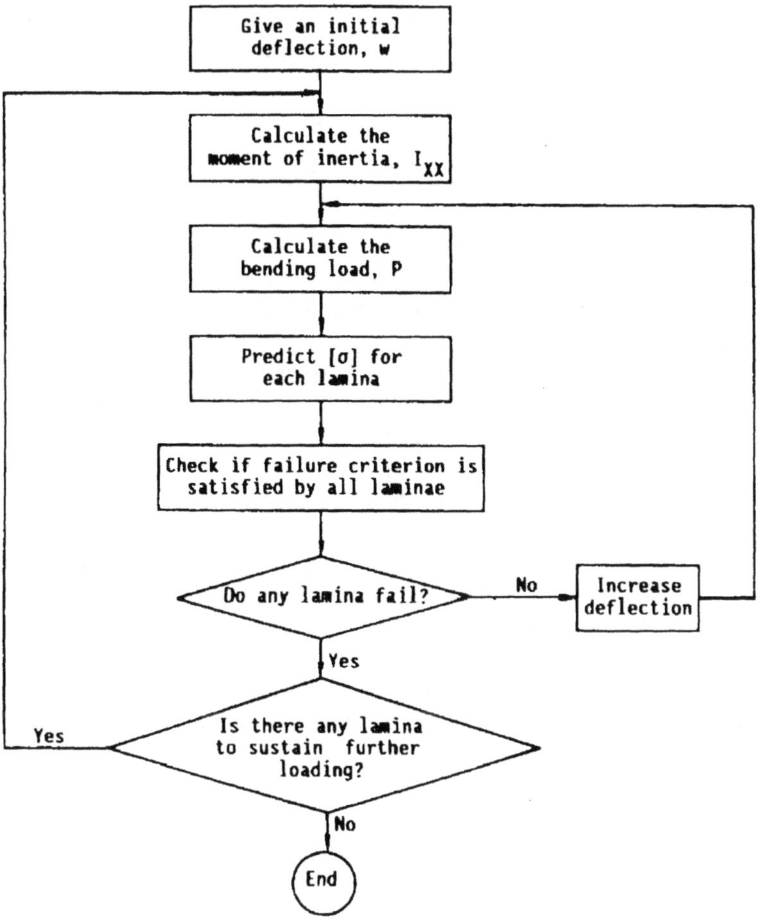

Figure 9.21. Flow-chart of the design procedure for a cantilever thin-walled beam subjected to bending.

$$w = \frac{PL^3}{3E_{XX}I_{XX}} \tag{9.17}$$

where, E_{XX} is the modulus of elasticity in the X-direction.

Thus, the calculation procedure can be described as in the flow-chart of Figure 9.21.

9.4.5 Crashworthy Capability: Concluding Remarks

The crashworthy behaviour of a non-conventional hourglass cross-section frame rail made of a glass fibre-vinylester composite, which has been designed for using it in the construction of the apron location of the car body in order to obtain a high degree crashworthy performance of the car, has been studied in bending both theoretically and experimentally. The bending crush strength and the modes of collapse of such rail beams were also investigated. Various factors, such as clamping end and loading conditions, affecting their structural ability and energy absorbing capacity, were examined and the obtained microscopic fracture patterns and the failure governing mechanisms were analysed in detail.

From the bending crush tests of this structural component, it was observed that the first fracture always occurred at the compressive zone, under the clamping nose, spreading towards the side walls and the clamped end, which is strongly distorted. The tensile zone remained intact.

The development of delamination and flexural cracks at the compression zone and the shear cracks in the form of shear bands at the side walls, the corners and the compression side constitute the main fracture mechanisms of the bending process.

The plug insert in each end of the tube plays an important role on the lengthening of the post-buckling regime, extending, therefore, the relevant M/θ diagrams to higher hinge rotation values, whilst the maximum strength is not affected by the plug insert. The experimental results were found to be in good agreement with the theoretical ones by applying the lamination theory combined by the Tsai-Wu failure criterion. The proposed analysis provides also the possibility to the designer to describe the M/θ curve in the elastic regime; see the various steps on the flow-chart in Figure 9.21.

Bending over the strong axis of the beam results in higher strength ability and better energy absorbing characteristics of the hourglass automotive section.

CHAPTER 10

CLASSIFICATION OF MACRO- AND MICROFAILURE MODES AND QUANTITATIVE DATA

10.1 COMMON DEFECTS IN THE PROCESSING OF COMPOSITE MATERIALS

Researchers into the mechanics of materials processing are mostly devoted to finding the load to perform an operation. This is indeed important for work-scheduling, i.e. deciding which machines are capable of applying any requisite force or power. However, there are occasions when other features may be even more important and deserving of attention. Such is the case concerning the defects and the limitations to processing, which can and do arise in products; they have tended to draw little attention from academic workers even though their economic consequences can be great.

The range of material working and fabrication defectiveness embraces [115]

- The occurrence of defects due to interaction between the workpiece material, the tooling, the friction between the latter and the process-geometry
- Some forms of microstructure, which result from purely mechanical action
- The limits of performance imposed by the material properties themselves with a given tooling and stressing system
- Elastic springback and generated residual stresses

Interaction of the above-mentioned features during meterial processing makes it difficult to account precisely for the defects met in terms of the mechanics; certain defects are associated with particular processes whilst some defects are peculiar to some materials.

A wide range of common principal defects that arise in composite material production and subsequent processing is presented in Reference [115]. The kinds of composites treated are:

- Fibre reinforced plastics
- Metal-matrix composites

- Clad/bonded materials
- Coated composites

In Table 10.1, the physical defects associated with composite material fabrication and processing as reported in Reference [115] are summarized. An extensive list of references is also provided in Reference [115].

10.2 COMMON DEFECTS IN LOADED COMPOSITE THIN-WALLED STRUCTURES

10.2.1 Axial Collapse: Static and Dynamic

The crush behaviour and the energy absorbing characteristics of fibre-reinforced structural components of various shell geometries, such as circular and square tubes, circular and square frusta and an automotive hourglass cross-sectioned rail frame, subjected to axial collapse, were examined both theoretically and experimentally. Detailed macro- and microstructural investigations on a variety of crushed shells, reported in Chapters 5–9, provide useful information and they contribute greatly to the understanding of the processes involved.

Thin-walled structures made from composite materials and subjected to axial collapse were found to collapse in modes considerably different than those observed in metallic and thermoplastic structures. The brittle nature of both fibres and resin ensures that composite materials do not undergo the characteristic for ductile metals and PVC plastic deformation. On the contrary, the mechanism of fracture and fragmentation dominates rather the crushing phenomenon. In general, the failure modes observed throughout the tests are greatly affected by the shell geometry, the arrangement of fibres, the properties of the matrix and fibres of the composite material and the stacking sequences. In Table 10.2 an attempt is made to classify the main macroscopic and microscopic collapse modes, pertaining to the axial collapse of composite thin walled structures. Their main features are outlined below.

MACROSCOPIC

Four modes of collapse at macroscopic level were observed, designated as Mode I, II, III and IV:

- *Mode I* (an end-crushing mode) is characterised by progressive collapse starting at one end of the shell. This collapse mode was observed in three different types of failure. The first one (Mode Ia) is mainly characterised by progressive collapse through the formation of continuous fronds, which spread outwards and inwards; it was observed in the case of tubes of various geometries and for circular and square frusta of small semi-apical angles. The second one (Mode Ib) is characterised by the collapse of the shell at the narrow end and the inversion of the tube wall inwards; it was observed for circular and square frusta

Table 10.1: Physical defects in composite material fabrication

Defect	Cause
FIBRE-REINFORCED PLASTICS	
Incomplete impregnation of fibre	Manufacture
Incomplete cure of resin	
Poor wetting and subsequent poor adhesion of fibre to matrix	
Bubbles	
Voids	
Delaminations	
Broken strands	
Loose ends of fibres	
Knotted strands	
Wrinkled strands and crevices	
Crazing cracks	
Local resin-rich areas	
Severe delamination	Forming
Concealed cuts	
Rupture of resin starved layers	
Fibre pull-out	Tensile loading
Fibre–matrix debonding	
Splitting	Compressive loading
Buckling	
Transverse cracking of fibres	Static and dynamic piercing
Parallel splitting of laminates	
Delamination	Impact
Translaminal cracking perpendicular to fibres	
Spalling	
Surface flaws (step, hole, ripple, branch, fissure, crack) of whiskers	
METAL–MATRIX COMPOSITES	
Incompatibility of fibre and matrix	Manufacture
Poor wettability	
Reaction between fibre and matrix	
Inadequate percolation of the matrix material to properly surround the fibres	
Voids	
Porosity	
Matrix–filament debonding	
Axial densification	Powder compaction
Density gradients	
Delaminations	
Filaments to break	
Filament buckles	
Formation and flattening of ribbons	
Flaky area of carbon fibre	Fibre-coating
Void formation at the poles of fibres	Forging
Transmatrix cracks	
Decohesion of a matrix–fibre interface	
Brooming or crushing of component ends	
Transverse cracking	Rolling
Rough 'tree-bark' surface finish	
Delaminations	
Fibre rupture as multiple necking	
Transverse cracking	Extrusion
Rough 'tree-bark' surface finish	
Delaminations	
Burst	
Split into bundles of metal-coated fibres	
Microcracks	
Bamboo shape defect	Drawing of bar, rod and wire
Breakage of continuous fibres or filament	
Voids	
Debonding	
Filament breakage	Bending
Splitting	
Brittle fracture	
SANDWICH, CLAD/BONDED MATERIALS	
Poor bonding	Manufacture
Bond breakage	
Warping at high temperatures	
Edge delamination	
Destruction of the flyer plate	
Wrinkling	Deep-drawing and bending
Earing	
Wall fracture	
Delamination or bowing	
Surface defects	Abrasive wear
COATED MATERIALS	
Poor coating	Manufacture
Thinning and chipping at the corners and edges	
Bubbles and specks due to excessive pickling	
Fish scale, ruptures, flakes, peeling off due to absorbed hydrogen between coating and base material	
Severe oxidation	
Substrate	
Voids	
Porosity	
Overspraying	
Surface marks	Sheet forming
Wrinkling	
Earing	
Springback	

Table 10.2: Classification of macro and micro failure modes and quantitive data for axially loaded composite thin-walled structures

Shell geometry	Material	Loading conditions	Collapse modes observed	Critical ratios		Main macro-characteristics of the collaped modes	Main micro-failures	Semi-apical angle (°)	Specific energy, W_s (kJ/kg)		(%) Percentage of dissipated energy to the main energy sources			
				$t/\overline{D}$ or t/C^*	$L/\overline{D}$ or L/C				Exper.	Theor.	Friction	Fronds bending	Crack propag.	Axial splitting
Circular tubes	A	Static	Ia			Ia: "mushrooming"	delaminations, fibre and matrix breakage	0	44.8	45.3	50	40	8	2
		Impact	Ia			Ia: "mushrooming"	delaminations, fibre and matrix breakage, matrix shattering	0	38.5	37.3	48	38	7	7
	B	Static	Ia			Ia: "mushrooming"	delaminations, fibres fracturing or buckling	0	55.0	55.5	55	38	4	3
		Impact	Ia			Ia: "mushrooming", extensive axial splitting	delaminations, fibre debonding	0	64.8	67.0	57	35	4	4
Square tubes	A	Static	Ia			Ia: "mushrooming", axial splitting at the four corners	delaminations, fibre and matrix breakage	0	40.6	41.5	48	44	6	2
		Impact	Ia			Ia: "mushrooming", axial splitting at the four corners	delaminations, fibre and matrix breakage, matrix shattering	0	32.9	33.4	46	43	7	4
	B	Static	Ia, III		1.6	Ia: "mushrooming", axial splitting at the four corners III: mid-height fracture	Ia: delaminations, fibres fracturing or buckling III: brittle fracture	0	48.5	50.3	52	42	4	2

Table 10.2 (cont.)

		Impact	Ia, III**		0.8	Ia: "mushrooming", axial splitting at the four corners III: mid-height fracture	Ia: delaminations, fibre debonding III: extensive brittle fracture	0	62.2	60.2	54	40	4	2
Circular frusta	A	Static	Ia, Ib, Ic, IIa, IIb, III, IV	0.02	1.5	Ia: "mushrooming" Ib: inversion of shell wall inwards Ic: inversion of shell wall outwards IIa: longitudinal crack IIb: circumferential crack III: mid-height fracture IV: sharp hinges	Ia: delaminations, fibre and matrix breakage Ib-Ic: extensive delaminations, successive shearing IIa-IIb-III: brittle fracturing IV: fracture hinges and folds	5	65.0	65.9	49	43	6	2
								10	50.2	51.7	48	46	4	2
								15	45.7	46.2	48	48	2	2
								20	15.3	19.0	30	45	25	-
		Impact	Ia, Ib	0.02	1.5	Ia: "mushrooming" Ib: inversion of shell wall inwards	Ia: delaminations, fibre and matrix breakage, matrix shattering Ib: delaminations, successive shear.	5	40.8	42.2	46	44	5	5
								10	34.2	37.5	45	47	3	5
								15	25.5	26.9	44	49	2	5
Square frusta	A	Static	Ia, Ib, II, III, IV	0.001	0.4	Ia: "mushrooming" Ib: inversion of shell wall inwards II: longitudinal cracks at the corners III: mid-height fracture IV: sharp hinges	Ia: delaminations, fibre and matrix breakage Ib: extensive delaminations, successive shearing II-III: brittle fract. IV: fracture hinges and folds	5	39.5	40.5	48	43	5	4
								10	40.5	41.2	47	45	4	4
								15	35.5	36.8	45.5	47	3.5	4
		Impact	Ia, Ib, II**	0.001	0.4	Ia: "mushrooming" Ib: inversion of shell wall inwards II: longitudinal	Ia: delaminations, fibre and matrix breakage, matrix Ib: delaminations, successive shear.	5	36.6	38.9	47	44	6	3
								10	34.2	34.6	45	47	5	3

Table 10.2 (cont.)

						cracks at the corners	II: extensive brittle fracturing	15	30.0	32.0	43	49	4	4
Auto-motive sections	B	Static	Ia, III		1.4	Ia: "mushrooming" III: mid-height fracture	Ia: delaminations, fibres fracturing or buckling III: brittle fracture		56.5	54.9	55	38	4	3
		Impact	Ia, III**		0.5	Ia: "mushrooming" III: mid-height fracture	Ia: delaminations, fibre debonding III: extensive brittle fracture		68.3	66.2	57	35	4	4
	B1	Static	Ia, III		0.8	Ia: "mushrooming" III: mid-height fracture	Ia: delaminations, fibres fracturing or buckling III: brittle fracture		49.5	51.2	55	35	6	4

* In the case of square tubes and frusta and automotive sections
** Complete catastrophic

of higher semi-apical angles. Finally, Mode Ic is characterised by the progressive inversion of tube wall outwards starting at the large end of the shell; it was observed only for 10° circular frusta.

- *Mode II* is a transition mode of collapse showing successive formations of an end-crushing mode, which is limited by shell failure, associated with the formation of a longitudinal or a circumferential crack at a just lower position in the case of circular frusta, or with the formation of cracks at the corners, which propagate towards the narrow end of the specimen, in the case of square frusta; it was observed for circular and square frusta of very high semi-apical angles (θ >20°).
- *Mode III* shows the characteristic features of brittle fracture involving catastrophic failure, which occurs at a distance from the frustum end approximately equal to the mid-height of the frustum; it was observed in the case of very high tubes with small wall thicknesses and circular and square frusta of small semi-apical angles.
- *Mode IV* constitutes the progressive folding mode and is characterised by the formation of sharp hinges; it was observed for very thin tubes and circular and square frusta with semi-apical angles not exceeding 15°. Thin-walled structures can dissipate large amount of energy by stable collapse (Mode I) when subjected to axial loading.

The characteristic modes of collapse, which are observed throughout dynamic axial tests, can be identified and classified as stable and unstable collapse modes. However, at impact, due to the dynamic nature of the phenomenon occured, unstable modes of collapse may lead rapidly to a complete catastrophic failure. It must be noted that, the stable modes obtained at impact loading show similar characteristics to those observed during the corresponding quasi-static ones.

The experimentally obtained deformation modes of all specimens tested may classified, in respect to the geometric ratios $t/\overline{D}$ (thickness/mean diameter of the shell) and $L/\overline{D}$ (length/mean diameter of the shell), for circular tubes and frusta and, in respect to the ratios t/C (thickness/circumference of the shell) and L/C (length/circumference of the shell), for square tubes and frusta. Distinct regions, characterising the various deformation modes occurred and the transition boundaries from one mode to another, are indicated. Shell instability occurs for values of the geometric factors lower than the critical ones, which are almost identical for static and dynamic loading, namely about 0.001 for t/C and 0.4 for L/C. Note that the critical value of L/C is almost equal to the related equivalent one for circular tubes and frusta subjected to similar loading conditions, whilst the value of the t/C is much higher for the square tubes and frusta as compared to the equivalent of circular ones.

MICROSCOPIC

The most common case of collapse, e.g. tubes of various geometries following the stable collapse mode, is characterised by the formation of a *central interlaminar opening crack* at the apex of a highly *pulverised wedge* of constant shape over its circumference, which separates the tube wall in two lamina-bundles of equal thickness. The debris wedge remains essentially unchanged during the compression process and penetrates the composite material, developing high frictional resistance with the adjacent fronds. The size of main crack is small compared to the tube axial length.

As far as the microfracture mechanism of circular and square frusta is concerned, it must be noted that in Mode Ia of collapse, the size of this wedge increases during crushing, since the diameter of the wedge increases as the cone crushing progresses. With increasing semi-apical angle of the frustum, the position of the intrawall crack moves towards the outside edge of the shell wall, increasing in this manner the thickness of the inner frond and, simultaneously, resulting in a positioning of the annular wedge, mainly above it. The dimensions of the triangular cross-section of the pulverised wedge and the crack length are greatly reduced for higher semi-apical angles and, therefore, during crushing, an internal and an external frond are formed without a bundle wedge formation, whilst a shear zone develops in both sides of the central crack. In Mode Ib, progressive collapse is created by successive shearing of the region near the narrow end of the shell.

Fracturing and/or *buckling* of the *reinforcing fibres* depends upon their orientation. Axially aligned fibres bend inwards or outwands, with or without fracturing, according to their flexibility and the constraints induced by other fibres; their effective flexibility depends upon the fibres structure in the composite material. Fibres aligned in the hoop direction can only expand outwards by fracturing and inwards by either fracturing or buckling.

Delamination occurs as a result of shear and tensile separation between plies. The axial laminae split into progressively thinner layers, forming, therefore, *translaminar cracks* normal to the fibres direction, mainly due to fibre buckling, finally resulting either in *fibre fracture* or in *intralaminar shear cracking*, which splits the laminate into many thin layers without fibre fracture. Cracks propagate preferably through the weakest regions of the structure of the composite material, i.e. through resin-rich regions or boundaries between hoop fibres, resulting in their debonding or through the interface between hoop and axial plies causing delamination.

The principal sources of energy dissipation at microscopic scale, which may contribute to the overall energy absorption during collapse, are: Intrawall crack propagation, fronds bending owing to delamination between plies, axial splitting between fronds, flexural damage of individual plies due to small radius of curvature at the delamination limits, frictional resistance to axial sliding between adjacent laminae, frictional resistance to the penetration of the annular debris wedge and frictional resistance to fronds sliding across the platen.

In general, the microfracture mechanism of Mode Ia of collapse is similar for statically and dynamically loaded shells. The only differences encountered are associated with the shape of the pulverised wedge and the microcracking development. The size of the debris wedge and the main central crack dimensions are smaller in the case of the impacted shells. During the dynamic loading, the resin behaved in a more brittle manner and it was shattered and separated almost completely from the fibres in the crush zone, an indication of the brittle behaviour of the composite material exhibited during its loading at elevated strain-rates. In the case of tubes, the wedge angle, α was estimated to about 90°, whilst in the case of the statically loaded specimens it was about 100°–110°. Also for the dynamically loaded circular and square frusta, the main cracking caused by the wedge is smaller in size whilst, the position of the wedge, which is smaller in size, and of the tip of the crack are shifted towards the outside wall surface of the impacted shells, as compared to statically loaded ones. Therefore, transition from Mode Ia to Mode Ib occurs for smaller semi-apical angles ($\theta = 10°$) for dynamical loaded frusta, as compared to the related static ones. From measurements of the collapsed impacted shells, the wedge angle was estimated to about 75°, 60° and 50° for 5°, 10°, and 15° frusta, respectively, whilst, in the case of statically loaded specimens, the wedge angle, α was estimated to about 80°, 70° and 60° for 5°, 10°, and 15° frusta, respectively.

ENERGY ABSORPTION: QUANTITATIVE DATA

For thin-walled composite shells, subjected to axial collapse, the fracture behaviour of the shell appears to affect the loading stability, as well as the magnitude of the crush load and the energy absorption during the crushing process; shells which collapse in a stable, progressive and controlled manner can dissipate a large amount of energy.

The energy absorption capability of a crushed structural component made of composite material is commonly quoted in the form of its specific energy absorption value, W_T, defined as the ratio of the energy absorbed, W for the collapsed specimen per unit mass crushed, m_C, calculated as the crushed volume, V_C times the material density, ρ. The obtained values of specific energy, for the shell geometries tested, tabulated in Table 10.2, were ranging from 20 up to 70 kJ/kg. From the results reported, it is indicated that the energy absorbing capability of the automotive sections and circular tubes is better (about 15–20%) than that of the square tubes made of the same material and tested under the same conditions, mainly due the shell geometry and the fracture mechanism associated with it. Also the increase of the specific energy for the circular frusta, statically loaded, is greater than for the corresponding square ones for all slenderness ratios, whilst for the dynamically ones, the specific energy seems to be slightly affected by the geometrical differences of the shells. From the energy absorbing capability point of view, conical shells of 5° semi-apical angle seem to be more efficient, due to the fact that an optimum combination of high values of energy dissipated, both for frictional work and fronds bending, was obtained and theoretically verified. The microfracture mechanism of this shell geometry is very similar to that of circular tubes.

Independent of geometry, all shells made of composite material A and subjected to dynamic loading, absorbed lower amounts of energy, about 20%, as compared with those absorbed in static collapse, probably due to the lower values of the dynamic friction coefficients between the wedge/fronds and fronds/platen interface and to the effect of the strain-rate on the microfracture mechanism. Contrariwise, in the case of shells made of composite material A, dynamic collapse overestimates static collapse by about 15%, probably due to the different values of the dynamic coefficients for material B, as compared to the related ones for material A. It seems that, the governing mechanism in this case is the friction mechanism than the changes in microfailures. Therefore, it may be concluded that the characteristics of the shell material greatly affect the crashworthy behaviour of shells at impact loading.

The mean post-crushing load, $\overline{P}$, the energy absorbed, W_T and the specific energy, W_s are well predicted theoretically by the proposed analysis within $\pm$10-15%. This theoretical approach is valid for all fibreglass composite materials, provided that the geometry and materials properties are known, but it gives better results for shells with even number of layers. However, in this approach, fibre orientation is not taken into account. For the prediction of the crushing loads and the absorbed energy of the dynamically loaded shells it is assumed that, the tensile fracture stress and the fracture toughness of the material remain constant, whilst these parameters can be a function of the crush-speed.

From the proposed analysis the distribution of the dissipated energy of the crushed tubes (circular and square tubes and automotive sections), associated with the main energy sources, can be estimated as: Energy absorbed, due to friction between the annular wedge and the fronds and between the fronds and the platen about 48–50%, due to bending of the fronds about 40–44%, due to crack propagation about 5–7% and due to axial splitting about 2–3%. As far as the Mode Ia of collapse of the conical shells (circular and square frusta) is concerned, from the proposed analysis, the distribution of the dissipated energy (average results for all the semi-apical angles used) of the crushed shell, associated with the main energy sources, can be estimated as: Energy absorbed, due to friction between the annular wedge and the fronds and between the fronds and the platen about 46–50%, due to bending of the fronds about 40–44%, due to crack propagation about 5–7% and due to axial splitting about 1–2%. It is, therefore, evident that the frictional conditions between wedge/fronds and fronds/platen constitute the most signifigant factors to the energy absorbing capability of the shell; the friction coefficients, μ_{s1}, μ_{d1} and μ_{s2}, μ_{d2}, are greatly affected by the surface conditions at the interfaces between composite material/platen or drop-mass and composite material/debris wedge, respectively.

In the case of conical shells following the Mode Ib of collapse the distribution of the energy absorbed due to friction, fronds bending and delaminations was estimated to about 30%, 45% and 25%.

The number of axial splits, which is affected by material properties, lay-up and shell geometry, is greater for the tubes of material B. In general, from the energy absorbing capacity point of view, shells of material B seem to be more efficient, as predicted both theoretically and experimentally; this is mainly due to the fabrication and the increased strength of the former material.

10.2.2 Bending

Since oblique impact and vehicle rollover are also common modes of automotive accidents, collapse of structural members due to bending becomes the predominant mode of crash energy absorption. Therefore, the bending of thin-walled fibre-reinforced composite shells, such as circular, square and rectangular tubes and automotive sections, under certain end-clamping conditions, simulating in this manner the oblique collision of structural elements of impacted vehicles, was both experimentally and theoretically investigated, see Chapters 5, 6 and 9.

MACRO/MICROSCOPIC

Experimental observations indicate that the predominant failure mode is an extensive microfracturing of the tubular component adjacent to the edge of the clamping device. The collapse mechanism of the bent composite material tubes is complex and strongly influenced by the type of fibres and matrix system employed and the nature and strength of the fibre-matrix interface bond. *Fracture of matrix* and *fibres*, *debonding of the fibres from the matrix* and friction required to pull broken fibres from the matrix must be taken into account.

Depending on shell geometry, two or three dinstict regions, with different macro- and microscopic characteristics were observed throughout the bending process of the tubes tested; the upper zone, mainly subjected to compressive loading, and the lower one under tensile straining. A narrow transition zone between them, with combined features, has been also observed. Note that, in the case of square and rectangular tubes, this transition zone is more profound since the side walls of the tubes are associated with combined compression/tension features and, therefore, it may be considered as a third zone. The fracture mechanisms governing the above mentioned zones are quite different.

Collapse initiates in the compressive zone, close enough to the clamping device. In this region, *fibre buckling* occurs and the strength of the structure is dominated by local fibre buckling stability, which depends on fibre diameter and modulus and on the support given by the matrix. Matrix compressive failure by slipping and/or fracturing is the result of the high stress field developed due to the associated forces.

In the tensile region, damage initiates and propagates in bands of high stress concentrations, such as *delamination edges* etc., where *micro-* and *macro-cracks* first develop in the matrix phase. This matrix damage, however, does not lead to catastrophic failure. On the contrary, fracture occurs only, when the fibre phase in the loading direction is sufficiently overstressed to reach the fibre fracture strength. Because of the broken fibres and the load release, their neighbouring fibres are loaded by shear through the matrix over a characteristic stressed length, leading, therefore, to stress concentrations in the fibre phase; this tends to be relieved by *matrix cracking*, parallel to the fibres, which in turn is stabilised by confinements introduced by the transverse plies.

The failure mechanism described above, which introduces significant strains perpendicular to the fibres, causes transverse cracking. The main features of this fracture are greatly affected by the fixture conditions at the clamped end of the tube.

In the case of circular tubes, a plug inserted into the inside diameter of the tube usually leads to full separation of the shell in two parts, with main characteristics the *pulled-out fibres*. Moreover, in cases of simple clamping, during the bending process, the upper shell region bends inwards, showing deformation characteristics similar to those observed for steel tubes.

Regions under compressive loading are mainly characterised by cracking, which in most specimens is developed at an angle of about 45° with reference to the tube axis. On the contrary, tube regions loaded in tension exhibit complex heterogeneous damage and failure modes, such as *matrix cracking, fibre/matrix interface debonding*, mainly for transverse fibres, and *delamination* and *fibre breakage*, mainly for longitudinal fibres.

The severity of damage depends on the laminar strength and stiffness, the stacking sequence, the ply orientation, the fibre volume fraction and the specimen size.

In the case of square/rectangular tubes, a typical progressive failure mechanism governs the bending process. It is characterised by an initial fracture along the longitudinal axis of symmetry of the flat of the compressive side of the tube, close to the clamping edge and, after spreading quickly from the centre towards the corners, it results in the development simultaneously of a large number of *delamination cracks* at the corners, just underneath the clamping device. As the hinge rotation continues, the tube sustains higher loads as the corners begin to crush, whilst its compressive side and side walls begin to buckle, finally leading to an overall separation of the tube wall. The extent of tube separation is more profound for thinner specimens.

From the bending crush tests of automotive sections, it was observed that, the first fracture always occurred in the compressive zone under the clamping nose, spreading towards the side walls and the clamped end, which is strongly distorted. The tensile zone remained intact. Moreover, the development of *delamination* and *flexural cracks* in the compression zone and *shear cracks* in the form of shear bands at the side walls, corners and the compression side of the shell are the main fracture mechanisms of the bending process.

ENERGY ABSORPTION: QUANTITATIVE DATA

The energy absorbing capability of the shell is calculated, in every case, by measuring the area under the corresponding bending moment, M/angle of rotation, θ curve. During the bending process, the shell initially deforms elastically and the M/θ curve is characterised by a sharp, steady-state increase of the bending moment until a maximum value, M_{max} is attained and cracking occurs. The post-crushing regime follows and it is characterised by deep collapse, where the initially developed transverse crack spreads gradually or rapidly over the whole cross section of the shell, mainly depending upon the clamping conditions.

In the case of circular tubes, the insertion of a plug as additional clamping device leads to higher initial maximum values of bending moment with simultaneous shortening of the post-crushing regime, probably due to the acceleration of the crack propagation, resulting in lower amounts of energy dissipated. The length of the plug plays also a role in the collapse efficiency (rigidity) of the bent tube, resulting in a significant increase of both the initial maximum bending moment and the energy absorbing capacity. Clamping devices with rounded edges delay the crack development and propagation, causing, in general, higher values of bending moment and amounts of energy absorbed.

As far as the energy absorbing capability of square and rectangular tubes subjected to bending is concerned, it must be noted that, the maximum bending moment seems to mainly depend upon the strength of the compressive top side and the buckling strength of the tube side walls. Furthermore, from the comparison of square/rectangular tubes with their equivalent circular tubes, it was observed that the circular tubes initially behave in a better manner, as far as absorption efficiency is concerned, whilst as the deformation progresses, the corners of the non-circular tubes play a significant role by improving the absorption efficiency, leading, therefore, to better crashworthiness characteristics for the latter.

Finally, in the case of automotive sections it was observed that, the plug insert in each end of the tube plays an important role on the lengthening of the post-buckling region, extending the relevant M/θ diagrams to higher hinge rotation values, whilst the maximum strength is not affected by the plug insert.

Bending over the strong axis of the beam shows higher strength ability and better energy absorbing capacity characteristics.

A theoretical analysis, for the prediction of the ultimate bending strength for thin-walled non-circular tubes subjected to bending, was presented. The maximum bending moment seems to mainly depend upon the strength of the compressive top side and the critical strength of the tube side walls and is well predicted theoretically by the proposed analysis (applying the lamination theory combined by the Tsai-Wu criterion). It increases with increasing wall thickness. The proposed analysis provides also the possibility to the designer to describe the M/θ curve in the elastic regime.

1 Johnson W. and Mamalis A.G. (1978), "Crashworthiness of vehicles", Mechanical Engineering Publications, London, U.K.

2 Johnson W. and Reid S.R. (1978), "Metallic energy dissipating systems", Appl. Mech. Rev. 31, 277.

3 Rawlings B. (1974), "Response of structures to dynamic loads", Proc. Conference on Mechanical properties at high rates of strain, The Institute of Physics, Oxford 279.

4 Franchini E. (1969), "The crash survival space", SAE Technical Paper Series, SAE 690005, New York, USA.

5 Ezra A.A and Fay R.J. (1972), "An assessment of energy absorbing devices for prospective use in aircraft impact situations", in "Dynamic response of structures", G. Hermann and N. Perrone Eds., Pergamon Press, 225.

6 Goldsmith W. and Sackman J.L. (1991), "Energy absorption by sandwich plates: A topic in crashworthiness", in "Crashworthiness and occupant protection in transportation systems, ASME, AMD, 126.

7 Johnson W., Mamalis A.G. and Reid S.R. (1982), "Aspects of car design and human injury", Ch. 4 in "Human body dynamics: impact, occupational and athletic aspects", D. N. Ghista, Eds., Clarendon Press, Oxford, 164.

8 Mamalis A.G., Manolakos D.E. and Viegelahn G.L. (1989), "Deformation characteristics of crashworthy components", Fortschritt-Berichte der VDI-Z, Reihc 18, Nr. 62, Dusseldorf, Germany.

9 Mamalis A.G., Robinson M., Manolakos D.E., Demosthenous G.A., Ioannidis M.B. and Carruthers J. (1997), "Crashworthy Capability of Composite Material Structures: A Review", Composite Structures, (in Press).

10 Morris C.J. (1990), "Taurus "TUB" all composite body structure demonstration vehicle", Proc. of the Sixth Annual ASM/ESD Advanced Composite Conference, Structural Composites (Design and Processing Technologies), Detroit, Michingan, USA, 53.

11 O'Rourke B.P. (1989), "The uses of composite materials in the design and manufacture of Formula 1 racing cars", Conference on Design in composite materials, Proc. Instn. Mech. Engrs., 39.

12 Savage G. (1992), "Safety and survivability in Formula One motor racing", Metals and Materials, (3), 147.

13 Walden D.C. (1990), "Applications of composites in commercial airplanes", Proc. 6th Annual ASM/ESD Advanced Composite Conference, Structural Composites (Design and Processing Technologies), Detroit, Michigan, USA, 77.

14 Lee S.M. (1993), "Handbook of composite reinforcements", VCH Publishers, Inc., USA.

15 Scholes A. and Lewis J.H. (1993), "Development of crashworthiness for railway vehicle structures", Proc. Instn. Mech. Engrs., Part F, 207, (F1), 1.

16 Lin K.H. and Mase G.T. (1990), "An assessment of add-on energy absorption devices for vehicle crashworthiness", Trans. ASME, J. Engineering Materials and Technology, 112, 406.

17 Tong P. (1983), "Rail vehicle structural crashworthiness", Ch. 14 in "Structural crashworthiness", N. Jones and T. Wierzbicki, Eds., Butterworths, London, 397.

18 Johnson W. (1990), "The elements of crashworthiness: Scope and actuality", Proc. Instn. Mech. Engrs., 204, (D4), 255.

19 Noton B.R. (1974), "Rail Transportation", Ch. 5 in "Composite Materials Vol. 3, "Engineering Applications of Composites", Academic Press, 163.

20 Brown D.E. (1994), "Choice of materials and form of construction of passenger vehicles", Proc. of the Eleventh Railway Industry Association Motive Power Course, RP3/21(1).

21 Suzuki Y. and Satoh K. (1995), "High speed trains and composite material", SAMPE Journal, September/October, 18.

22 Thornton P.H. and Jeryan R.A. (1988), "Crash energy management in composite automotive structures", Int. J. Impact Engineering 7, 167.

23 Granta Design Limited (1994), "The Cambridge Materials Selector Version 2.0", Software package running on an IBM compartible PC.

24 Ashby M.F. (1992), "Materials selection in mechanical design", Pergamon Press.

25 McCarty J.E. (1993), "Design and cost viability of composites in commercial aircraft", Composites 24, 361.

26 Frame T.S. (1989), "Introduction to composite materials", Proc. Conference on Design in composite materials, Proc. Instn. Mech. Engrs, 1.

27 Agarwal B.D. and Broutman L.J. (1990) "Analysis and performance of fiber composites", John Wiley & Sons Inc, New York, USA.

28 Rosen B.W. (1985), "Tensile failure of fibrous composites", AIAA J. 2, (11).

29 Chamis C.C., Hanson M.P. and Stefanini T.T., "Designing for impact resistance with unidirectional fiber composites", NASA TN D-6463.

30 Broutman L. J. (1965), "Glass-resin joint strengths and their effect on failure mechanisms in reinforced plastics", Modern Plast.

31 Greszezuk L.B. (1974), "Microbuckling failure of circular fiber reinforced composites", Proc. 15th Structure, Structural Dynamics and Materials Conference, AIAA/ASME/SAE, USA.

32 Coolings T.A. (1974), "Transverse compressive behaviour of unidirectional carbon fibre reinforced plastics", Composites 5, 108.

33 Mamalis A.G., Manolakos D.E. and Demosthenous G.A. (1992), "Crushing behaviour of thin-walled non-circular fibreglass composite tubular components due to bending", Composites 23, 425.

34 Carlsson L.A. and Pipes R.B. (1987), "Experimental characterization of advanced composite materials", Prentice-Hall, Englewood Cliffs, N.J., USA.

35 Witney J.M., Daniel I.M. and Pipes R.B. (1981) " Experimental mechanics for fiber reinforced composite materials, SESA Monograph No. 4, Society of Experimental Stress Analysis, Brookfield Center, Conn., USA.

36 Hofer K.E. and Rao P.N. (1977), "A new static compression fixture for advanced composite materials", J. Testing Eval., 5.

37 Rosen B.W. (1972), "A simple procedure for experimental determination of the longitudinal shear modulus of unidirectional composites", J. Composite Materials 6.

38 Pindera M.J. and Herakovich C.T. (1986), "Shear characterization of unidirectional composites with off-axis tension test", Experimental Mechanics 26 (1).

39 Chamis C.C. and Sinclair J.H. (1977), "Ten-degree off-axis test for shear properties in fiber composites", Experimental Mechanics 17 (9).

40 Johnson W. (1972), "Impact strength of materials", Arnold, London, U.K.

41 Yuan Y.B. and Viegelahn G.L. (1991), "Modelling of crushing behaviour fibreglass/vinylester tubes", Proc. 22nd Midwestern Mechanics Conference, Rolla, Missouri, USA, "Developments in Mechanics", 16, 490.

42 Pipes R.B. and Pagano N.J. (1970), "Interlaminar stresses in composite laminates under uniform axial extension", J. Composite. Materials. 4, 538.

43 Wilkins D.J., Eisenmann J.R., Camin R.A, Margolis W.S. and Benson R.A. (1980), "Characterizing delamination growth in graphite-epoxy", Damage in Composite Materials, ASTM STP 775, 168.

44 Anderson G.P., Bennett S.J. and DeVries K.L. (1977), "Analysis and testing of adhesive bonds", Academy Press, N.Y., USA.

45 Bascom W.D., Bitner R.J., Moulton R.J. and Siebert A.R. (1980), "The interlaminar fracture of organic-matrix woven reinforced composites", Composites 11, 9.

46 Kendall K. (1976), " Interfacial cracking of a composite". J. Material Science 11, 638 & 1263.

47 Nahas M.N. (1986), "Survey of failure and post-failure theories of laminated fiber-reinforced composites", J. Comp. Tech. Res., 8, (4).

48 Jones R. (1975), "Mechanics of composite materials", Scripta Book Company, USA.

49 Vinson J. and Chou T.W. (1975), "Composite materials and their use in structures", Applied Science Publishers Ltd, London, U.K.

50 Hill R. (1950), "The mathematical theory of plasticity", Oxford University Press, Oxford, U.K.

51 Tsai S.W. (1968), "Strength theories of filamentary structures", Ch. 1 in "Fundamental aspects of fiber reinforced plastic composites", R.T. Schwartz and H.S. Schwartz, Eds., Interscience, New York, USA.

52 Tsai S.W. and Wu E.M. (1971), "A general theory of strength for anisotropic materials", J. Composite Materials 5, 58.

53 Tsai S.W. and Hahn H.T. (1980), "Introduction to composite materials", Lancaster, PA: Technomic, USA.

54 DeRouvray A. and Haug E. (1989), "Failure of brittle and composite materials by numerical methods", Ch. 7, in "Structural Failure", T. Wierzbicki and N. Jones, Eds., Jo. Wiley & Sons, New York, USA.

55 PAM-FISS, A specialized Finite Element Code for Fracture Mechanics Analyses, Ref. 4185, ESI-SA, Rungis-Cendex, France.

56 Haug E. and DeRouvray A. (1984), "Toughness criterion identification for edge notched unidirectional composite laminates", Joint Inst. Adv. Flight Sci. (JIAFS), NASA, Georges Washington University, Arlington, VA.

57 Haug E., DeRouvray A. and Dowlatyari P. (1985), "Delamination criterion identification for multilayered composite laminates", Composite Structures, 20.

58 Prinz R. (1983), "Growth of delamination under fatigue loading", Proc. 56th Structure and Materials Panel Meeting of AGARD, London, U.K., DFVLR paper.

59 DeRouvray A., Vogel F. and Haug E. (1986), "Investigation of micromechanics for composites", Phase 1 Rep. Vol. 2, ESI Rep. ED/84-477/RD/MS (under ESA/ESTEC contract), Eng. Sys. Int., Rungis-Cendex, France.

60 DeRouvray A., Dowlatyari P. and Haug E. (1987), "Investigation of micromechanics for composites", Phase 2a Rep. WP2, ESI Rep. ED/85-521/RD/MS (under ESA/ESTEC contract), Eng. Sys. Int., Rungis-Cendex, France.

61 Haug E., Fort O. and Tramecon A. (1991), "Numerical crashworthiness simulation of automotive structures and components made of continuous fiber reinforced composite and sandwich assemblies", SAE Technical Paper Series, SAE, 910152, 245.

62 Mamalis A.G., Manolakos D.E., Viegelahn G.L. and Baldoukas A.K. (1990), "Bending of fibre-reinforced composite thin-walled tubes", Composites 21, 431.

63 Thornton P.H. (1979), "Energy absorption in composite structures", J. Composite Materials 13, 247.

64 Farley G.L. (1983), "Energy absorption of composite materials", J. of Composite Materials 17, 167.

65 Schmueser D.W. and Wickliffe L.E. (1987), "Impact energy absorption of continuous fiber composite tubes", J. Engineering Materials and Technology Trans. ASME 72, 72.

66 Farley G.L and Jones R.M. (1992), " Crushing characteristics of continuous fibre-reinforced composite tubes", J. Composite Materials, 26, 37.

67 Farley G.L and Jones R.M. (1992), "Analogy for the effect of material and geometrical variables on energy absorption capability of composite tubes", J. Composite Materials 26, 78.

68 Thornton P.H. and Edwards P.J. (1982), "Energy absorption in composite tubes", J. Composite Materials 16, 521.

69 Farley G.L. (1986), "Effect of fibre and matrix maximum strain rate on the energy absorption of composite materials", J. Composite Materials, 20, 322.

70 "PE fibre reinforcement prevents crush", News article in British Plastics and Rubber, January (1990).

71 Hamada H., Coppola J.C., Hull D., Maekawa Z. and Sato H. (1992), "Comparison of energy absorption of carbon-epoxy and carbon-PEEK composite tubes", Composites 23, 245.

72 Gosnell R.B. (1987), "Thermoplastic resins", Engineered Materials Handbook, Vol. 1- Composite, ASM International, 97.

73 Nilson S. (1991), "Polyetheretherketone matrix resins and composites", International Encyclopaedia of Composites, Vol. 6, VCH Publishers Inc., 282.

74 Kindervater C.M. (1990), "Energy absorption of composites as an aspect of aircraft structural crash resistance", in "Developments in the Science and Technology of composite materials", 643.

75 Mamalis A.G., Manolakos D.E., Demosthenous G.A. and Ioannidis M.B (1996), "Analysis of failure mechanisms observed in axial collapse of thin-walled circular fibreglass composite tubes", Thin-Walled Structures 24, 335.

76 Mamalis A.G., Manolakos D.E., Demosthenous G.A. and Ioannidis M.B (1996), "The static and dynamic axial crumbling of thin-walled fibreglass composite square tubes", Composites Engineering 6, (to be published).

77 Mamalis A.G., Manolakos D.E., Demosthenous G.A. and Ioannidis M.B. (1996), "The static and dynamic collapse of fibreglass composite automotive frame rails", Composite Structures 34, 77.

78 Farley G.L. (1986), "Effect of specimen geometry on the energy absorption of composite materials", J. Composite Materials 20, 390.

79 Farley G.L and Jones R.M. (1992), "Crushing characteristics of composite tubes with "near-elliptical" cross sections", J. Composite Materials 26, 1252.

80 Mamalis A.G., Yuan Y.B. and Viegelahn G.L. (1992), "Collapse of thin-wall composite sections subjected to high speed axial loading", In. J. Vehicle Design 13, 564.

81 Czaplicki M.J., Robertson R.E. and Thornton P.H. (1991), "Comparison of bevel and tulip triggered pultruded tubes for energy absorption", Composites Science and Technology 40, 31.

82 Mamalis A.G., Manolakos D.E., Demosthenous G.A. and Ioannidis M.B. (1994), "Axial collapse of thin-walled fibreglass composite tubular components at elevated strain rates", Composites Engineering 4, 371.

83 Mamalis A.G., Manolakos D.E., Viegelahn G.L., Demosthenous G.A. and Yap S.M. (1991), "On the axial crumpling of fibre-reinforced composite thin-walled conical shells", Int. J. Vehicle Design 12, 450.

84 Farley G.L. (1991), " The effect of crushing speed on the energy-absorption capability of composite tubes ", J. Composite Materials 25, 1314.

85 Kirch P.A. and Jannie H.A. (1981), "Energy absorption of glass polyester structures", SAE Technical Paper Series, SAE 810233, 220.

86 Berry J. and Hull D. (1984), "Effect of speed on progressive crushing of epoxy-glass cloth tubes", Proc. 3rd Int. Conference on Mechanical Properties at High Rates of Strain, Institute of Physics Series Oxford, 463, U.K.

87 Thornton P.H. (1990), "The crush behaviour of pultruded tubes at high strain rates", J. Composite Materials 24, 594.

88 Mamalis A.G., Manolakos D.E., Demosthenous G.A. and Ioannidis M.B (1994), "On the bending of automotive fibre-reinforced composite thin-walled structures", Composites 25, 47.

89 Mahmood H.F., Jeryan R.A. and Zhou J. (1990), "Crush strength characteristics of composite structures", Proc. of the Sixth Annual ASM/ESD Advanced Composite Conference, Structural Composites (Design and Processing Technologies), Detroit, Michingan, USA, 1.

90 Czaplicki M.J., Robertson R.E. and Thornton P.H. (1990), "Non-axial crushing of E-glass/polyester pultruded tubes", J. Composite materials 24, 1077.

91 Mamalis A.G., Manolakos D.E. and Viegelahn G.L. (1990), "Crashworthy behaviour of thin-walled tubes of fibreglass composite material subjected to axial loading", J. Composite Materials 24, 72.

92 Mamalis A.G., Manolakos D.E., Demosthenous G.A. and Ioannidis M.B. (1996), "Energy absorption capability of fibreglass composite square frusta subjected to static and dynamic axial collapse", Thin-Walled Structures, 25, 269.

93 Fairfull A.H. and Hull D. (1988), " Energy absorption of polymer matrix composite structures: Friction effects", chap. 8 in "Structural Failure", T. Wierzbicki and N. Jones, Eds., J. Wiley & Sons, New York, USA, 255.

94 Mamalis A.G., Manolakos D.E., Viegelahn G.L., Yap S.M. and Demosthenous G.A. (1991), "Microscopic failure of thin-walled fibre-reinforced composite frusta under static axial collapse", Int. J. Vehicle Design 12, 557.

95 Mamalis A.G., Manolakos D.E., Demosthenous G.A. and Ioannidis M.B. (1996) " Analytical modelling of the static and dynamic axial collapse of thin-walled fibreglass composite conical shells", Int. J. of Impact Engineering, (in press).

96 Mamalis A.G., Manolakos D.E., Demosthenous G.A. and Ioannidis M.B (1995), "The deformation mechanism of thin-walled non-circular composite tubes subjected to bending ", Composite Structures 30, 131.

97 Farley G.L and Jones R.M. (1992), "Prediction of the energy absorption capability of composite types", J. Composite Materials 26, 338.

98 Botkin M.E., Johnson N.L., Halloquist J.O., Lum L.C.K. and Matzemiller A. (1994), "Numerical simulation of post-failure dynamic crushing of composite tubes", Proc. Second International LS-DYNA3D Conference, San Francisco, USA.

99 Yuan Y.B., Viegelahn G.L. and Mamalis A.G. (1991), "Crushing characteristics of composite sections subjected to high speed loading", Proc. 22nd Midwestern Mechanics Conference, Rolla, Missouri, USA, "Developments in Mechanics", 16, 450.

100 Viegelahn G.L., Johnson D.J., Wlosinski R.K. and Mamalis A.G. (1991), "Crashworthy characteristics of thin-walled tubes in bending", Proc. 22nd Midwestern Mechanics Conference, Rolla, Missouri, USA, "Developments in Mechanics", 16, 470.

101 Viegelahn G.L., Gorrepati K.M., Johnson D.J., Wlosinski R.K. and Mamalis A.G. (1991), "Crashworthiness characteristics of composite tubes in bending", Proc. 7th Annual ASM/ESD Advanced Composite Conference, "Advanced Composite Materials: New Developments and Applications", Detroit, USA, 481.

102 Mamalis A.G., Manolakos D.E., Baldoukas A.K. and Viegelahn G.L. (1989), "Deformation characteristics of crashworthy thin-walled steel tubes subjected to bending", Proc. Institution of Mechanical Engineers, Part C, J. Mechanical Engineering Science, 203, 411.

103 Hull D. (1982), "Energy absorption of composite materials under crash conditions", Proc. ICCM-IV, "Progress in Science and Engineering of Composites" T. Hayashi, K. Kawata and S. Umekawa, Eds. Tokyo, 861.

104 Wierzbicki T. and Abramowicz W. (1988), "Development and implementation of special elements for crash analysis", Proc. 7th Int. Conference on Vehicle Structural Mechanics, 113.

105 Tabor D. (1981), "Friction—The present state of our understanding", J. Lubrication Technology 103, 169.

106 Mamalis A.G. and Johnson W. (1983), "The quasi-static crumpling of thin-walled circular cylinders and frusta under axial compression", Int. J. of Mechanical Sciences 25, 783.

107 Mamalis A.G., Johnson W. and Viegelahn G.L. (1984), "The crumpling of steel thin-walled tubes and frusta under axial compression at elevated strain-rates: Some experimental results", Int. Journal of Mechanical Sciences 26, 537.

108 Mamalis A.G., Manolakos D.E, Viegelahn G.L., Vaxevanidis N.M. and Johnson W. (1986), "On the inextensional axial collapse of thin PVC conical shells", Int. J. of Mechanical Sciences 28, 323.

109 Hull D. (1983), "Axial crushing of fiber reinforced composite tubes", Ch. in "Structural Crashworthiness", N. Jones and T. Wierzbicki, Ėds., Butterworths, London, 118.

110 Mamalis A.G., Manolakos D.E., Demosthenous G.A. and Ioannidis M.B. (1996), "Experimental determination of splitting in axially collapsed thick-walled fibre-reinforced composite frusta", Thin-Walled Structures, 25 (to be published).

111 Price J. N. and Hull D. (1987), "Axial crushing of glass-polyester composite cones", Compos. Sci. Technol., 28, 211.

112 Mamalis A.G., Manolakos D.E., Demosthenous G.A. and Ioannidis M.B (1994), "Crashworthy characteristics of fibreglass composite automotive frame rails subjected to axial collapse and bending", Proc. 27th ISATA Conf. "New and Alternative Materials for the Transportation Industries", Aachen, Germany, 371.

113 Mamalis A.G., Manolakos D.E., Demosthenous G.A. and Ioannidis M.B. (1995), "Analytical and experimental approach to damage and residual strength of fibreglass composite automotive frame rails during manufacturing", Proc. Eighth International Conference on Composite Structures, Paisley, Scotland, Composite Structures, 32, 325.

114 Thornton P.H. and Magee C.L. (1977) "The interplay of geometric and materials variables in energy absorption", Trans. ASME, J. Engineering Materials and Technology, 99, 114.

115 Thornton P.H., Harwood J.J. and Beardmore P. (1985), "Fibre-reinforced plastic composites for energy absorption purposes", Comp. Sci. Tech., 24, 75.

116 Johnson W. and Mamalis A.G. (1984), "Common defects in the processing of metals and composite materials", "Plasticity Today" Conference, CISM, Udine, Italy, 1983, Applied Science Publishers, London, U.K.

AUTHOR INDEX

Note: Page numbers in bold type indicate references at the end of the book.

Milton Keynes UK
Ingram Content Group UK Ltd.
UKHW040108071024
449327UK00019B/913

9 780367 400354